Graph Theory

A Problem Oriented Approach

Hardcover version issued in 2008.
Paperback version issued in 2011.

Library of Congress Catalog Card Number 2010927260

Paperback edition ISBN: 978-0-88385-775-5

Electronic edition ISBN: 978-0-88385-969-2

Printed in the United States of America

Current Printing (last digit):
10 9 8 7 6 5 4 3 2

Graph Theory

A Problem Oriented Approach

Daniel A. Marcus

Published and distributed by
The Mathematical Association of America

MAA TEXTBOOKS

Calculus Deconstructed: A Second Course in First-Year Calculus, Zbigniew H. Nitecki
Combinatorics: A Guided Tour, David R. Mazur
Combinatorics: A Problem Oriented Approach, Daniel A. Marcus
Complex Numbers and Geometry, Liang-shin Hahn
A Course in Mathematical Modeling, Douglas Mooney and Randall Swift
Cryptological Mathematics, Robert Edward Lewand
Differential Geometry and its Applications, John Oprea
Elementary Cryptanalysis, Abraham Sinkov
Elementary Mathematical Models, Dan Kalman
Essentials of Mathematics, Margie Hale
Field Theory and its Classical Problems, Charles Hadlock
Fourier Series, Rajendra Bhatia
Game Theory and Strategy, Philip D. Straffin
Geometry Revisited, H. S. M. Coxeter and S. L. Greitzer
Graph Theory: A Problem Oriented Approach, Daniel Marcus
Knot Theory, Charles Livingston
Lie Groups: A Problem-Oriented Introduction via Matrix Groups, Harriet Pollatsek
Mathematical Connections: A Companion for Teachers and Others, Al Cuoco
Mathematical Interest Theory, Second Edition, Leslie Jane Federer Vaaler and James W. Daniel
Mathematical Modeling in the Environment, Charles Hadlock
Mathematics for Business Decisions Part 1: Probability and Simulation (electronic textbook), Richard B. Thompson and Christopher G. Lamoureux
Mathematics for Business Decisions Part 2: Calculus and Optimization (electronic textbook), Richard B. Thompson and Christopher G. Lamoureux
The Mathematics of Games and Gambling, Edward Packel
Math Through the Ages, William Berlinghoff and Fernando Gouvea
Noncommutative Rings, I. N. Herstein
Non-Euclidean Geometry, H. S. M. Coxeter
Number Theory Through Inquiry, David C. Marshall, Edward Odell, and Michael Starbird
A Primer of Real Functions, Ralph P. Boas
A Radical Approach to Real Analysis, 2nd edition, David M. Bressoud
Real Infinite Series, Daniel D. Bonar and Michael Khoury, Jr.
Topology Now!, Robert Messer and Philip Straffin
Understanding our Quantitative World, Janet Andersen and Todd Swanson

MAA Service Center
P.O. Box 91112
Washington, DC 20090-1112
1-800-331-1MAA FAX: 1-301-206-9789

For Shelley

Preface

This book developed from a course in graph theory that I have taught at California State Polytechnic University since 1984. My classes consist primarily of mathematics and computer science majors with a smaller number of engineering and other science students. The class level is generally third and fourth year, but no particular prerequisite is needed. I tell students to just bring their brains.

The format is similar to that of the companion text *Combinatorics, a Problem Oriented Approach* (MAA, 1998) in that it combines features of a traditional textbook with those of a problem book. The material is presented through a series of approximately 430 problems with connecting text and is supplemented by approximately 300 additional problems for homework assignments. The problems are arranged strategically to introduce concepts in a logical order and in a provocative way. My lectures usually consist of working problems at the board with class input, but there are other possibilities: Students might work problems in groups or someone might be assigned a series of problems to present, in effect delivering part of the lecture.

The book is organized in seventeen chapters, each covering a different topic. Each chapter is divided into two groups of problems, roughly identifiable as new material and homework problems, with the latter group beginning at the heading More Problems. At the end of the book are answers to selected problems from the first group.

Reader involvement

The problem oriented format is intended to promote active involvement by the reader while always providing clear direction. This approach figures prominently in the presentation of proofs, which become more frequent and elaborate as the book progresses. Arguments are arranged in digestible chunks and always appear along with concrete examples to keep the reader firmly grounded in the motivation. See, for example, proofs of the following:

Tree Theorem #1 in chapter D, starting with "Pruning a Tree";

The Path/Cycle Principle for the existence of a Hamilton cycle, beginning after problem G21;

The Five Color Theorem, beginning with problem K23;

The König-Egervary Theorem on independent sets in a bipartite graph, Dilworth's Theorem, following problem O16.

This book was published with the generous assistance of

Dr. Alan C. Krinik
Professor of Mathematics
California State Polytechnic University, Pomona.

Contents

Preface ix

Introduction: Problems of Graph Theory 1

Path Problems 1
Coloring Problems 2
Isomorphic Graphs 4
Planar Graphs 4
Disjoint Paths 5
Shortest Paths 6
... and More 7

A Basic Concepts 9

Equivalent Graphs 9
Multigraphs 10
Directed Graphs and Mixed Graphs 11
Complete Graphs 11
Cycle Graphs 11
Paths in a Graph 11
Open and Closed Paths; Cycles 12
Subgraphs 13
The Complement of a Graph 13
Degrees of Vertices 13
The Degree Sequence of a Graph 14
Regular Graphs 15
Connected and Disconnected Graphs 15
Components of a Graph 15
More Problems 16

B Isomorphic Graphs 21

More Problems 23

C Bipartite Graphs **25**
Complete Bipartite Graphs 26
Bipartite Graphs and Matrices 26
Cycles in a Bipartite Graph 27
Cycle Theorem for Bipartite Graphs 27
Proof of the Cycle Theorem 28
More Problems 29

D Trees and Forests **31**
Pruning a Tree 32
Directed Trees 33
Spanning Trees 34
Counting Spanning Trees 34
Codewords for Trees: Prufer's Method 35
More Problems 36

E Spanning Tree Algorithms **41**
Constructing Spanning Trees 41
Weighted Graphs 41
Minimal Spanning Trees 42
Prim's Algorithm 42
Tables for Prim's Algorithm 43
The Reduction Algorithm 43
Spanning Trees and Shortest Paths 44
Minimal Paths in a Weighted Graph 45
Minimal Path Algorithm, first attempt 45
Minimal Path Algorithm, revised 46
Tables for Dijkstra's Algorithm 46
Minimal Paths in a Directed Graph 48
Negative Weights 48
More Problems 51

F Euler Paths **57**
The Königsberg Bridge Problem 57
Euler Paths in Directed Graphs and Directed Multigraphs 59
Application of Euler Paths: State diagrams,
DeBruijn sequences, and rotating wheels 60
More Problems 61

G Hamilton Paths and Cycles **65**
Some Negative Tests 65
Positive Tests for Hamilton Cycles 67
Some Proofs 70
More Problems 73

H Planar Graphs **77**
Regions Formed by a Plane Diagram 78
Proof that $\mathbf{K}_5$ is Non-Planar, Using Euler's Formula 80

Non-Planar Graphs and Kuratowski's Theorem 81
More Problems 83

I Independence and Covering **85**
The Independence Numbers of a Graph 85
A Graph Game 88
Covering Sets and Covering Numbers 89
More Problems 91

J Connections and Obstructions **95**
Internally Disjoint Paths 95
Edge-Disjoint Paths 95
Path Connection Numbers 96
Blocking Sets 96
k-Connected Graphs 98
Vertex Cut Sets and Vertex Cut Numbers 99
More Problems 100

K Vertex Coloring **103**
The Vertex Coloring Number of a Graph 103
Vertex Coloring Theorems 104
Map Coloring 110
More Problems 111

L Edge Coloring **119**
The Edge Coloring Number of a Graph 119
Edge Coloring of Complete Graphs 120
Edge Coloring of Bipartite Graphs 122
Edge Color Switch 122
Proof of Edge Coloring Theorem #3 123
Application of Edge Coloring: the Scheduling Problem 124
More Problems 124

M Matching Theory for Bipartite Graphs **131**
The Max/Min Principle 132
Proof of the König–Egervary Theorem 133
The Colored Digraph Construction 133
Matching Extension Algorithm 135
Proof of the Colored Digraph Theorem 135
Matrix Interpretation of the König–Egervary Theorem 136
Hall's Theorem and Its Consequences 138
More Problems 139

N Applications of Matching Theory **143**
Sets and Representatives 143
Latin Squares 144
Permutation Matrices 145

The Optimal Assignment Problem . . . 146
More Problems . . . 149

O Cycle-Free Digraphs **153**
Chains and Antichains . . . 153
Chain Decompositions . . . 154
Proof of Dilworth's Theorem . . . 156
More Problems . . . 158

P Network Flow Theory **161**
Flows in a Network . . . 161
Cuts and Capacities . . . 169
More Problems . . . 172

Q Flow Problems with Lower Bounds **175**
The Supply and Demand Problem . . . 175
More Problems . . . 184

Answers to Selected Problems **187**

Index **201**

About the Author **205**

Introduction
Problems of Graph Theory

Path Problems

Is it possible to trace this figure in one continuous motion? The key to solving this problem lies in knowing where to start.

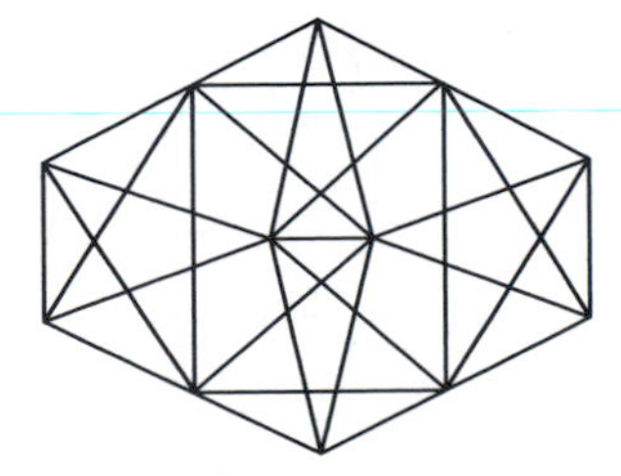

Closely related is the *Königsberg Bridge Problem*. In the city of Königsberg there were seven bridges connecting two islands and portions of the city on each bank of a river.

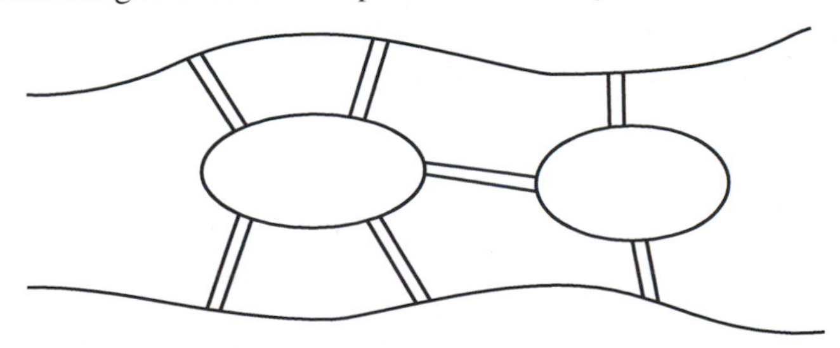

The residents tried to find a route that would cross each bridge exactly once. But they never succeeded, and for a good reason: They were trying to do the impossible. The explanation, along with the secret to tracing the figure in the first problem, appears in chapter F, where problems of this type will be viewed in the context of graph theory.

In general, a *graph* consists of points, which are called *vertices* (singular: *vertex*), and connections, which are called *edges* and which are indicated by line segments or curves joining certain pairs of vertices. Vertices that are joined by edges are called *adjacent vertices*. For example, the graph below contains 7 vertices and 15 edges.

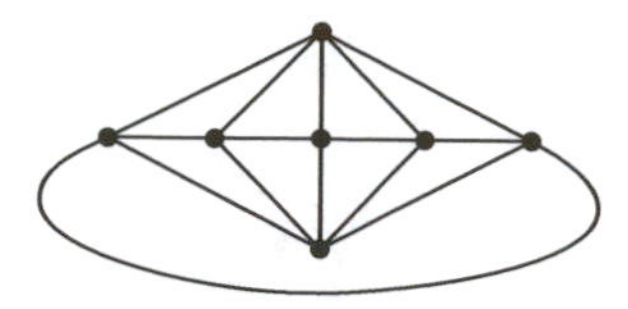

An *Euler path* in a graph is a route that covers every edge exactly once. See if you can find an Euler path in the example above.

A slightly different problem is to find a route in a given graph that visits every vertex exactly once but need not cover every edge. Such a route is called a *Hamilton path*. For example, it is easy enough to find a Hamilton path in this graph. Try it.

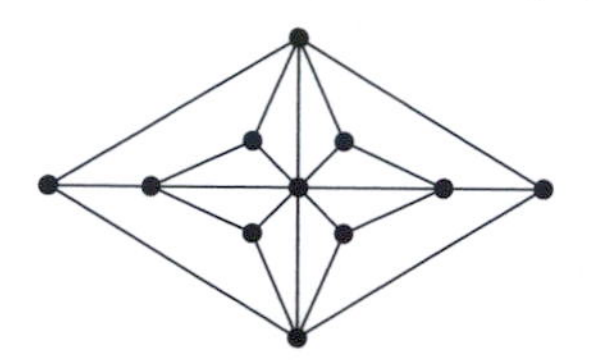

The problem becomes harder if we require a route that visits every vertex exactly once and then returns to the starting point. In other words, we want the final vertex of the path to be adjacent to the first. In that case we are looking for a *Hamilton cycle*. Can you find one in the example above? How about in this graph?

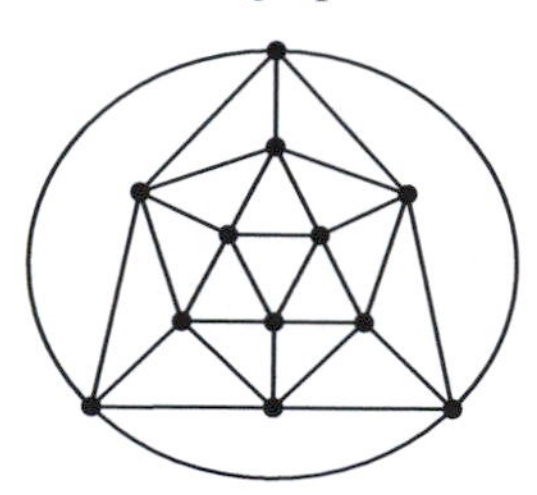

Hamilton paths and cycles will be discussed in chapter G. Among other things, we will see that in some graphs it is possible to decide that no Hamilton path or cycle exists without conducting an exhaustive search, while in certain others we will see how to predict that a Hamilton path or cycle exists without actually finding it.

Coloring Problems

Below is the map of a rectangular continent that contains nine countries.

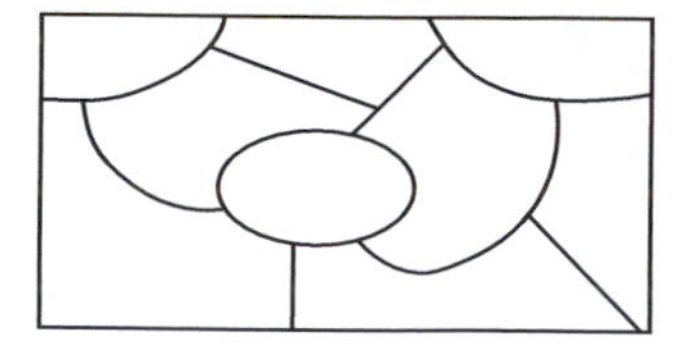

In coloring the map, it is reasonable to require that different colors be used for countries that share a common border. In that case, how many different colors are actually needed to color this map?

If you used more than four colors for the map above, go back and try again. The famous **Four Color Theorem** tells us that any map, no matter how large and complicated, can be colored using no more than four colors. This phenomenon was first observed at least 150 years ago but not proved until 1976 by a method that relied on extensive checking of details by computer. In chapter K and also later in this introduction we will see how the map coloring problem can be restated in terms of coloring the vertices of an appropriate graph. A *proper vertex coloring* of a graph is the assignment of a color (usually represented by a number or some other symbol) to each vertex of the graph in such a way that no color occurs at adjacent vertices. For example, a proper vertex coloring of the graph below is shown using five colors: red, blue, green, yellow and white.

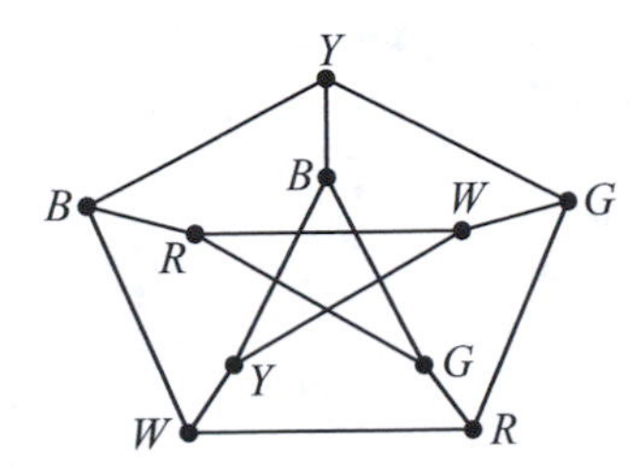

However, are all these colors really needed? Try to find a proper vertex coloring of this graph using the fewest possible colors.

A variation on the problem above is that of coloring the edges of a given graph. A *proper edge coloring* of a graph is the assignment of a color to each edge of the graph in such a way that no color occurs on edges that share a common endpoint. For example, here is a proper edge coloring of a graph using four colors.

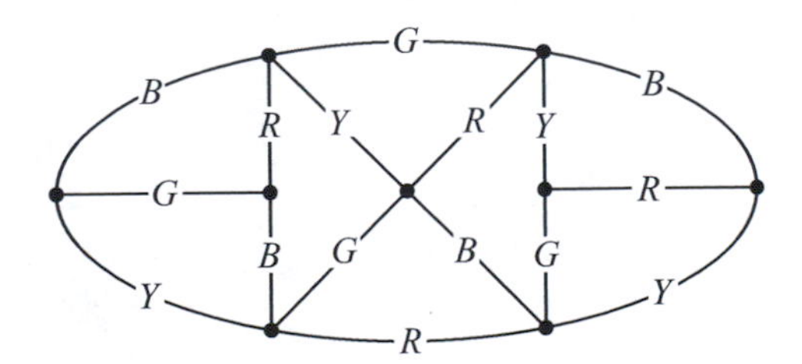

Can you see why it would be impossible to reduce the number of colors in this case? Try to find a proper edge coloring of the graph below using the fewest possible colors.

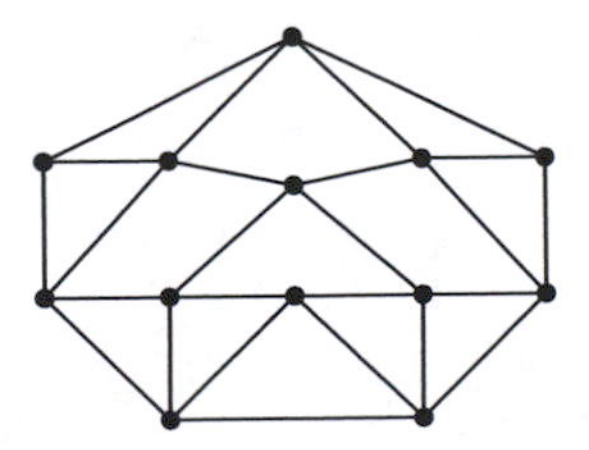

Edge coloring has many interesting applications, some of which are presented in chapters L and N. For example, can you place eight rooks on the blank spaces of the chessboard below in such a way that no rook can take any other? In other words, no two rooks are in the same row or column.

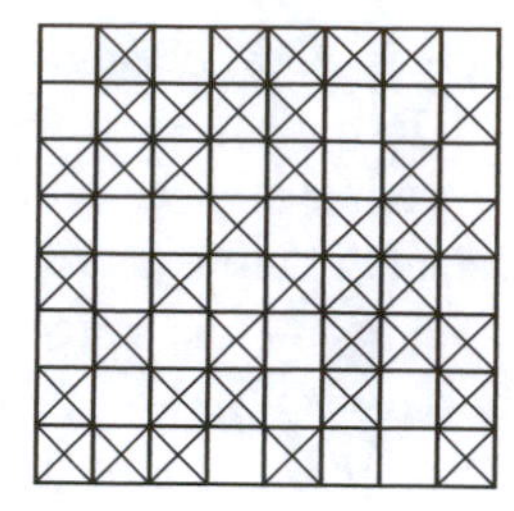

While this is not hard to do, it may surprise you that the same would be true on a square board of any size as long as each row and column contains the same number of blank spaces. We will see how this fact can be proved as a consequence of a theorem on coloring the edges in a certain type of graph.

Isomorphic Graphs

An important concept in graph theory is that of *isomorphic graphs*. For example, the two graphs below are essentially the same even though they appear different, in the sense that one can be transformed into the other by moving vertices around and changing the lengths of some of the edges that join them. It helps to think of the graphs as being made up of flexible wires that can be stretched or contracted.

Can you see why these two graphs are not isomorphic?

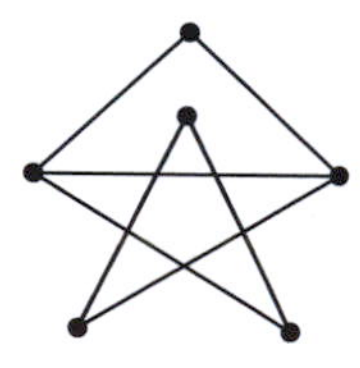

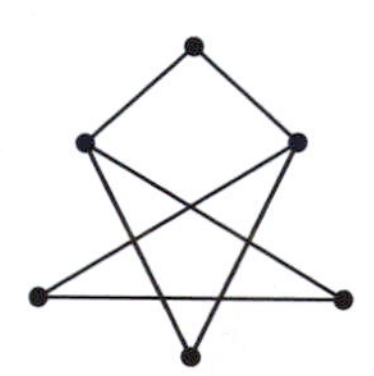

On the other hand, one of the above is isomorphic to this one. Which is it?

Planar Graphs

By now you have noticed that certain graphs contain edges that cross while others don't. A graph in which no edges cross is called a *planar graph*. The same term is also applied to any graph that can be redrawn in such a way that no edges cross. For example,

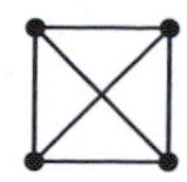

qualifies as a planar graph since it is isomorphic to this one:

But what about this graph?

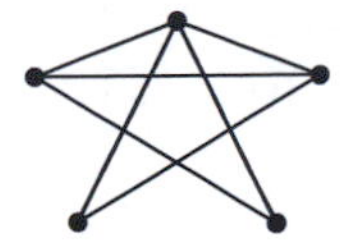

If you find a way to redraw it with no edges crossing, try making the problem harder by adding another edge at the bottom.

Planar graphs have an interesting application in solid geometry. For example, count the vertices (corners) and the faces (flat surfaces) in each of the figures below, and in each case compare the total of these two with the number of edges in the figure.

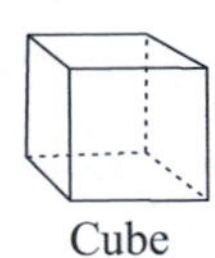

Cube

Tetrahedron

Hexahedron

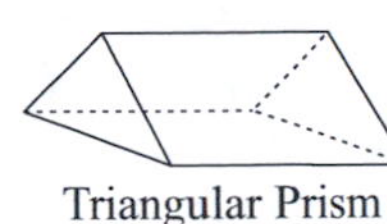

Triangular Prism

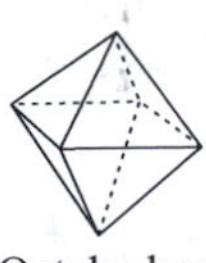

Octahedron

The difference is always the same (try it) and remains the same no matter what figure is used. In other words, **Euler's Formula** $v - e + f = 2$ holds for every 3-dimensional solid that has flat surfaces. This will be proved in chapter H with the help of planar graphs.

Planar graphs are also involved in the map coloring problem in an important way. The Four Color Theorem is equivalent to the statement that every planar graph has a proper vertex coloring that uses no more than four colors. The connection is suggested by the hint shown below.

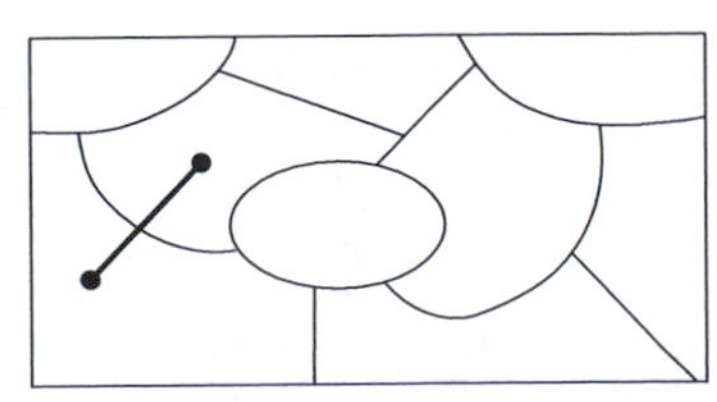

Disjoint Paths

Returning to path problems, see how many different paths you can find in this graph that go from A to B and have no common vertices except for their two endpoints. In other words, how many people can walk from A to B without taking paths that cross?

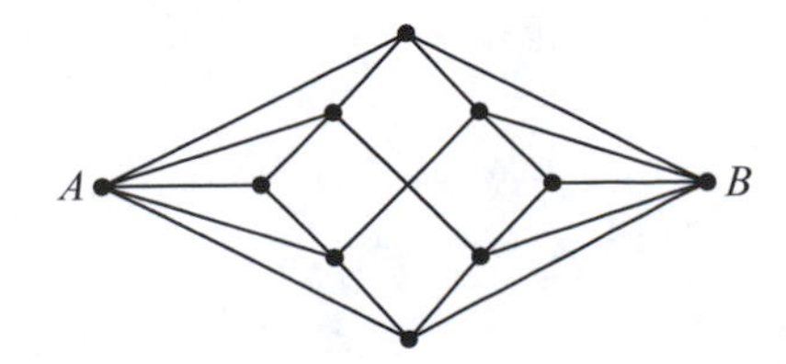

Such paths are called *internally disjoint*. Can you think of a good reason why there could not be five internally disjoint paths from A to B? Hint: Look for four points that all of these paths would have to cross.

On the other hand, suppose we allow paths to have common vertices but no common edges. These are called *edge disjoint* paths. Can five people all take edge disjoint paths from A to B?

Problems that require disjoint paths are dealt with in chapter J.

See how many internally disjoint paths and how many edge disjoint paths you can find going from A to B in this example:

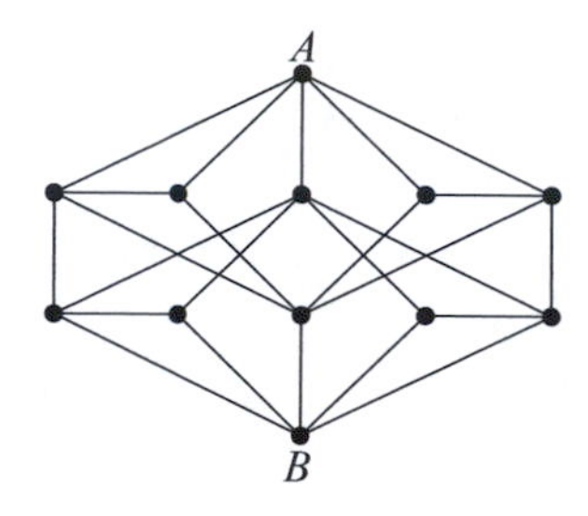

Shortest Paths

Finding a shortest route between two vertices in a given graph is an important problem with obvious real-life applications. For example, you can easily find a shortest path (one with fewest edges) going from A to B in this graph.

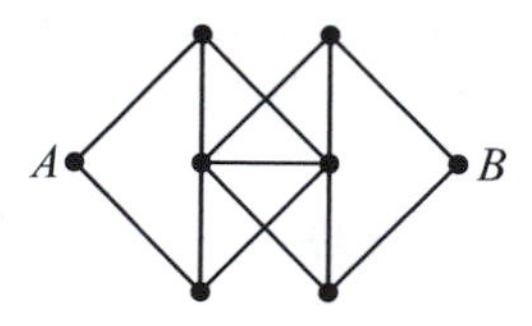

However suppose some of the edges (think of them as streets) have one-way signs. How does that change the result?

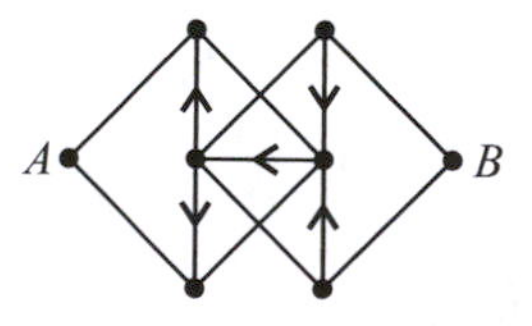

In chapter E we will give an algorithm for constructing a shortest path between given points in any graph, and we'll see how it works in more complicated situations where not all edges are considered equal. For example, the numbers in the graph below can be thought of as costs. Think of a toll booth on each edge.

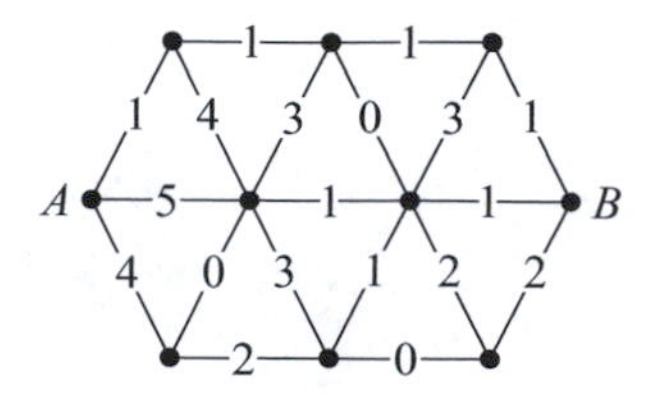

Each edge in a given path contributes to the total cost of the path, and the goal is to find a path between two given points that has the lowest possible cost. What would that be in this case if the given points are A and B?

... and More

Further applications of graphs appear in chapters N and O. Some of these may be surprising in that it is not immediately obvious why the problem has anything to do with a graph. For example, below is an example of an **optimal assignment problem** in which workers are rated for particular jobs.

	Bill	Ross	George	Al	Ralph
used car salesman	8	10	4	5	0
fraternity president	9	2	8	4	3
standup comic	7	4	1	5	9
talk show host	8	1	3	2	5

The goal is to fill the jobs in the best possible way, maximizing the sum of all ratings involved in the assignment. How would you assign these jobs?

In another application, we will see how a graph can help to determine whether it is possible to add a column to this matrix, using only the letters A through H, in such a way that no letter appears more than once in any row or column.

C	B	D	G	F	A
H	E	G	F	A	C
G	F	A	E	B	D
D	G	F	H	C	B
B	D	C	A	E	H
F	A	H	B	D	G
E	C	B	D	G	F

Can you do it? Or do you see a reason why it's impossible?

Finally, a *monotone sequence* of numbers is a sequence that is either increasing or decreasing, not necessarily strictly. For example, 3, 2, 2, 1 is a monotone sequence. What can the following problem possibly have to do with graph theory?

Let $a_1, a_2, \ldots, a_{101}$ be any sequence of 101 real numbers. Prove that the sequence contains a monotone subsequence of length 11.

If you read to the end of the book, you will find out how this problem is solved with the help of a graph.

Enough. It's time to get started.

Basic Concepts

A *graph* consists of points, which are called *vertices* (singular: *vertex*), and connections, which are called *edges* and which are indicated by line segments or curves joining certain pairs of vertices. In the graph below there are five vertices A, B, C, D, and E.

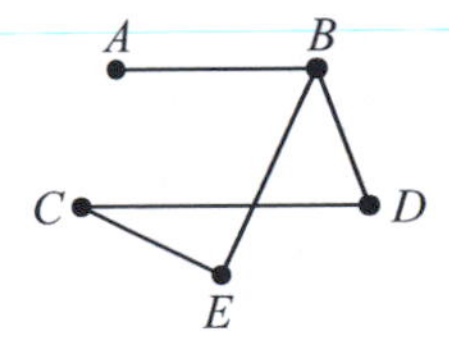

The pairs A and B, B and D, B and E, C and D, C and E, and E and D are joined by edges which are denoted as AB, BD, BE, CD, CE, and ED. Vertices that are joined by edges are called *adjacent vertices*. In this graph, A and B are adjacent vertices while A and C are nonadjacent.

All graphs in this book will be assumed to be *finite graphs*, which means that they contain a finite number of vertices and edges.

Equivalent Graphs

In graph theory, unlike in geometry, the shape of an edge or the position of the vertices is considered unimportant. Therefore these two graphs are considered to be the same, or *equivalent graphs*.

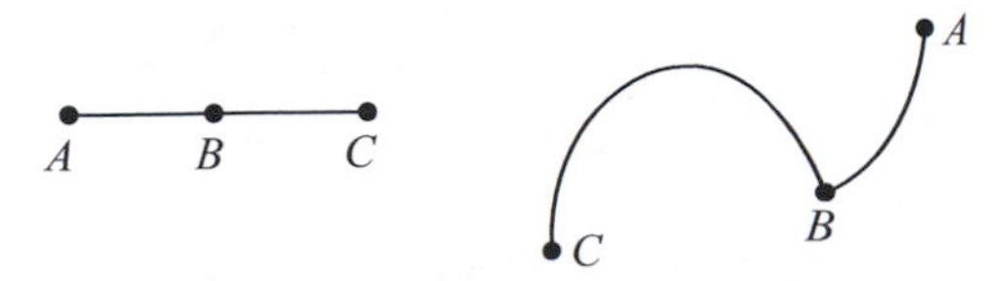

This means that in both graphs, the same pairs of vertices are adjacent.

A1 Of the six graphs shown below, which ones are equivalent to each other?

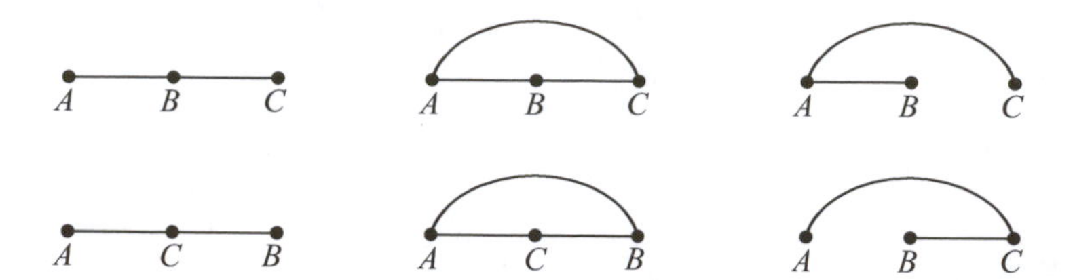

The formal definition of a graph is a set of elements, along with a set of two-element subsets of that set. The first set represents the vertices of the graph, while the second set represents the edges. For example, the graph

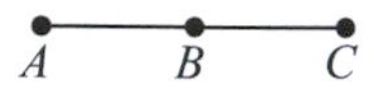

is formally represented by the two sets $\{A, B, C\}$ and $\{\{A, B\}, \{B, C\}\}$.

A2 Write the formal representation for this graph.

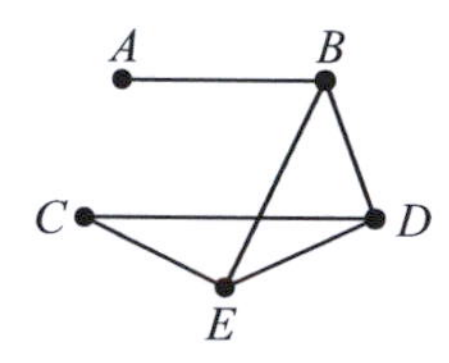

In all of the graphs so far, the vertices have been given names, or labels. These labels can be letters, numbers, or any other symbols. Also, sometimes the vertices of a graph will not be labeled at all. Therefore we make a distinction between *labeled graphs* and *unlabeled graphs*. This distinction will become important in Chapter B.

Edges in a graph are allowed to cross each other without intersecting at a vertex. Therefore the two graphs below are significantly different. The first graph contains four vertices and only two edges, while the second graph contains five vertices and four edges.

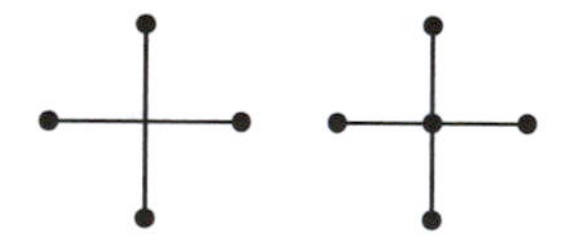

Multigraphs

Situations such as these will not be allowed in a graph:

We will require that every edge in a graph has two different endpoints (there are no *loops*) and that there is at most one edge joining any two vertices (there are no *multiple edges*, which are edges having the same two endpoints). The term *multigraph* is used when loops and/or multiple edges are allowed to occur.

Directed Graphs and Mixed Graphs

Generally, the edges in a graph will not have directions associated with them. However in a *directed graph*, or *digraph*, each edge includes a direction from one endpoint to the other. In a *mixed graph*, both directed and undirected edges are allowed.

Complete Graphs

If all of the vertices in a graph are adjacent to each other, then the graph is called a *complete graph*. The symbol $\mathbf{K}_n$ is used to denote a complete graph with n vertices. Some complete graphs are shown here:

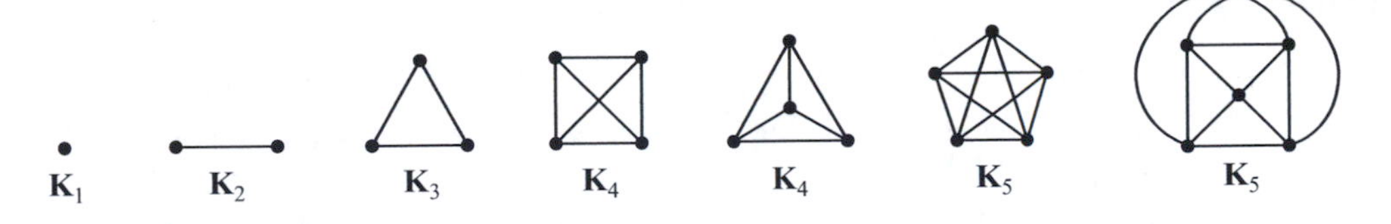

Cycle Graphs

For now, we will introduce *cycle graphs* by examples rather than by a formal definition. The idea is reasonably obvious: Think of a graph that can be completely traced by starting at one vertex and continuing on to new vertices until returning to the first vertex. A cycle graph containing n vertices will be denoted by the symbol $\mathbf{C}_n$.

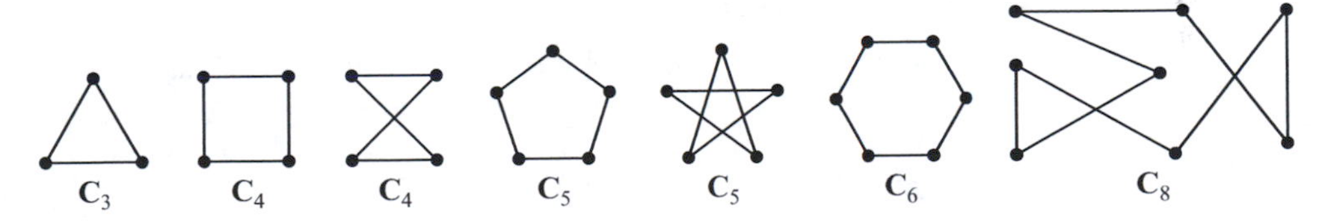

Paths in a Graph

A *path* is a sequence of vertices and edges in a graph such that

1. The sequence alternates between vertices and edges, starting and ending with vertices; and
2. Each edge in the sequence joins the vertices that occur immediately before and after it in the sequence.

For example, (A, AB, B, BC, C, CE, E) is a path in this graph starting at A and ending at E.

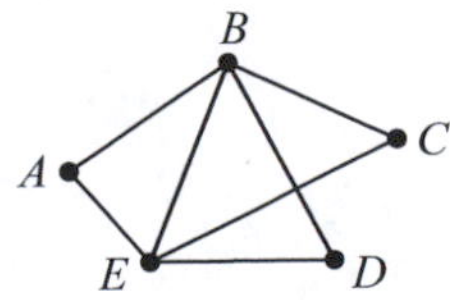

We will refer to this path as a *path from A to E*. Notice that the same path can be described more efficiently by leaving out the edges and just writing (A, B, C, E) or, even more simply, $ABCE$. (Why would we not be able to do this in a multigraph?)

Another path from A to E in this same graph is $(A, AB, B, BC, C, CB, B, BE, E)$, or $ABCBE$.

As this last example shows, repetitions of vertices and edges are allowed in a path. A path that contains no repetitions is called a *simple path*. For example, in the graph above, $ABCED$ is a simple path. However the sequence $ABCD$ is not a path at all. (Why not?)

The *length* of a path is defined as the number of edges in the path, including repetitions when they occur. In the example above, the path $ABCBE$ has length 4.

A3 (a) Find a shortest path (a path of minimum length) from A to B in this graph. Is your path a simple path?

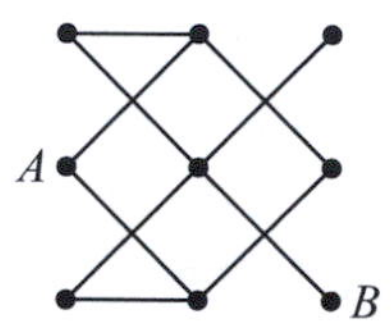

(b) In the same graph, find a simple path from A to B that is not a shortest path from A to B.

(c) Explain the relationship between simple paths and shortest paths joining two given vertices in a graph.

A4 (a) Find a longest simple path from A to B in the graph in problem A3.

(b) Why is it not possible to talk about a longest path joining two given vertices in a graph?

Open and Closed Paths; Cycles

An *open path* starts and ends at different vertices. A *closed path* starts and ends at the same vertex.

A5 Can a simple path in a graph also be a closed path?

A6 Make up a reasonable definition for a cycle graph. Start by saying that it consists of a closed path. What else has to be true?

In addition to cycle graphs, we can talk about cycles that are contained within a larger graph.

Definition A *cycle* in a graph is a closed path in which the only repetition is the first and last vertex.

A7 (a) Find three cycles of length 3 in this graph such that each cycle includes a different set of vertices.

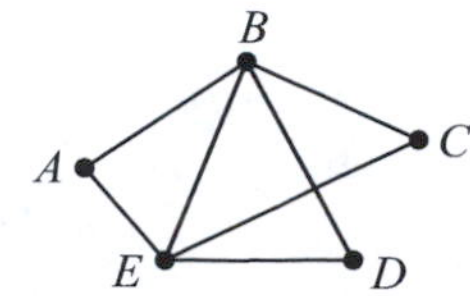

(b) In the same graph, find three cycles of length 4 such that each includes a different set of vertices.

(c) Explain how you know that there is no cycle of length 5 in this graph.

Subgraphs

A *subgraph* of a graph **G** is a graph that is contained within **G**. All vertices and edges of the subgraph must be included in **G**. For example, if this graph is **G**,

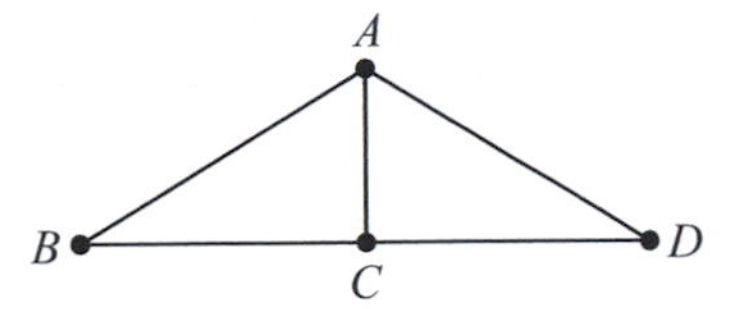

then these are some subgraphs of **G**:

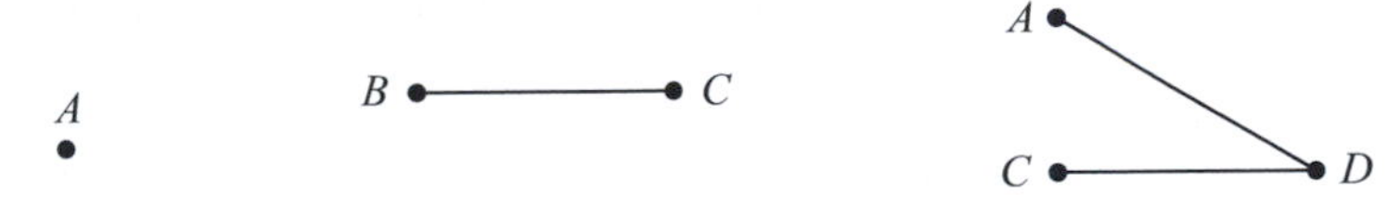

However this one is *not* a subgraph of **G**:

Also, any graph is considered to be a subgraph of itself.

The Complement of a Graph

The complement of a graph $\mathbf{G}_1$ is another graph $\mathbf{G}_2$ having the same set of vertices but including only those edges that are not in $\mathbf{G}_1$.

Example These two graphs are complements of each other.

Degrees of Vertices

Definition The *degree* of a vertex in a graph is the number of edges that occur at that vertex. The notation $d(A)$ represents the degree of vertex A.

A8 What is the degree of each vertex in a cycle graph $\mathbf{C}_n$? In a complete graph $\mathbf{K}_n$? In a graph that consists of just a simple path?

A9 Assume that $\mathbf{G}_1$ and $\mathbf{G}_2$ are complementary graphs, each with n vertices. Let A be one of these vertices. What relationship exists between the degree of A in $\mathbf{G}_1$ and the degree of A in $\mathbf{G}_2$? Refer to these numbers as $d_1(A)$ and $d_2(A)$.

A10 Add up the degrees of all of the vertices in this graph.

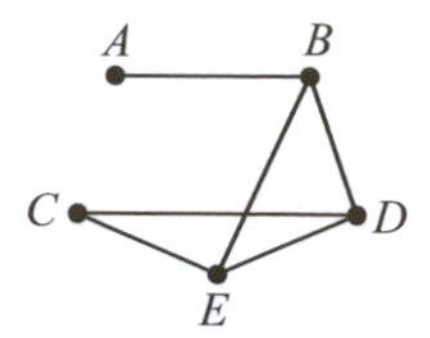

Then notice the number of edges in the graph. Guess what relationship might exist between the degree sum and the number of edges in any graph.

The Degree Theorem *In any graph, the sum of all the degrees is equal to twice the number of edges.*

A11 Explain why this is always true.

Definition An *odd* vertex in a graph is a vertex whose degree is an odd number. An *even* vertex is a vertex whose degree is an even number. Note that zero is an even number, so a vertex with no edges is considered an even vertex.

A12 Try to find a graph that has exactly one odd vertex. Then use the Degree Theorem to explain why this is impossible.

A13 How many odd vertices can a graph contain? What are the possibilities?

Notation In any graph **G**, the symbol δ, or $\delta(\mathbf{G})$, represents the minimum degree in **G**; Δ, or $\Delta(\mathbf{G})$, represents the maximum degree.

A14 (a) In a graph with eight vertices, what, if anything, can you say about the maximum degree Δ?

(b) In a graph with n vertices, what can you say about Δ?

A15 Suppose you know that a certain graph has seven vertices, and that $\delta = 3$ and $\Delta = 5$.

(a) Show that this graph must contain at least 12 edges. (Suggestion: use the Degree Theorem.)

(b) What is the largest number of edges possible in this graph?

The Degree Sequence of a Graph

The *degree sequence* of a graph is a list, or sequence, of all of the degrees of the vertices in the graph, including any repetitions. Usually we will list these degrees in decreasing order. So for example, the degree sequence of the graph in problem A10 is (3, 3, 3, 2, 1).

A16 Find a graph that has the degree sequence (3, 3, 2, 2, 2, 1, 1).

A17 For each sequence of numbers below, explain why there cannot be any graph having that degree sequence.

(a) (5, 5, 5, 4, 4, 3, 3)

(b) (7, 6, 5, 4, 3, 2, 1)

(c) (6, 5, 4, 3, 2, 2, 0)

Regular Graphs

A graph is called *regular* if all of its vertices have the same degree. If d is a nonnegative integer, then a d*-regular graph* is a graph in which each vertex has degree d.

A18 (a) Find a 1-regular graph that has 8 vertices.
(b) Describe all possible 1-regular graphs.

A19 Find a 2-regular graph that has n vertices.

A20 (a) Find a 3-regular graph that has 10 vertices.
(b) Explain why there cannot exist a 3-regular graph with 11 vertices.

Connected and Disconnected Graphs

Definition A graph is *connected* if every vertex is joined to every other vertex by a path. A *disconnected* graph is a graph that is not connected.

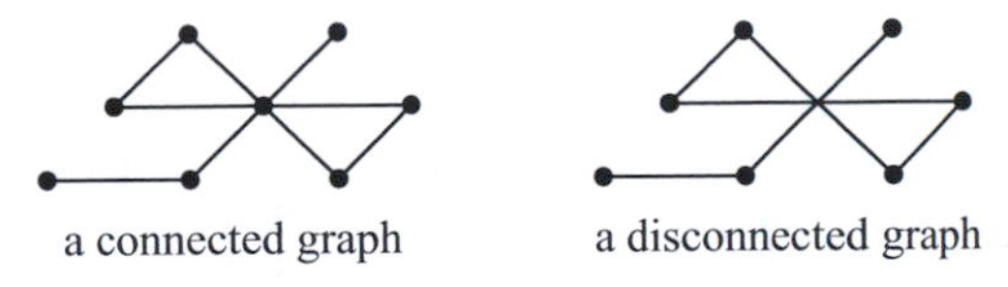

a connected graph a disconnected graph

A21 Is this graph connected or disconnected?

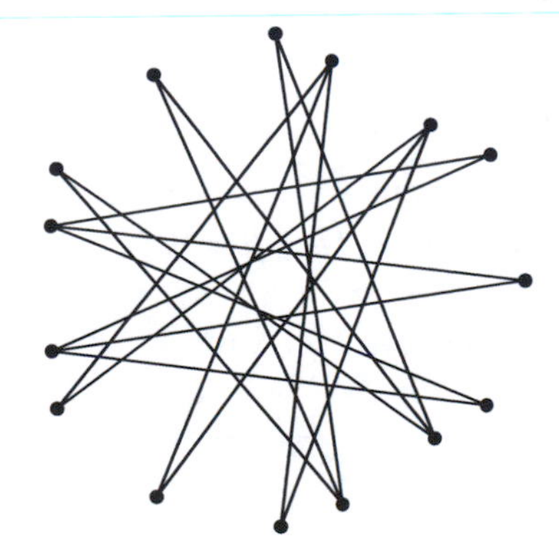

Components of a Graph

If a graph is disconnected, then it consists of two or more *components*. Each component is a connected subgraph **H** whose vertices are not adjacent to any vertices not in **H**.

Examples This graph has two components.

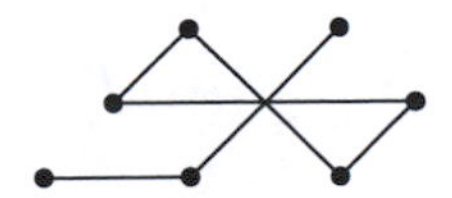

The graph in problem A21 has three components.

Think of the components of a graph as "connected pieces" of the graph that are disconnected from each other. A connected graph has only one component, which is the entire graph.

A22 Suppose that A and B are two vertices in a given graph.

(a) If A and B are adjacent, then are they necessarily in the same component of the graph?

(b) If A and B are in the same component of the graph, then are they necessarily adjacent?

(c) Fill in the blank appropriately: Two vertices A and B are in the same component of a graph if and only if there is a ______________.

More Problems

A23 In each case below, label the vertices of the second graph in such a way that the two graphs are equivalent.

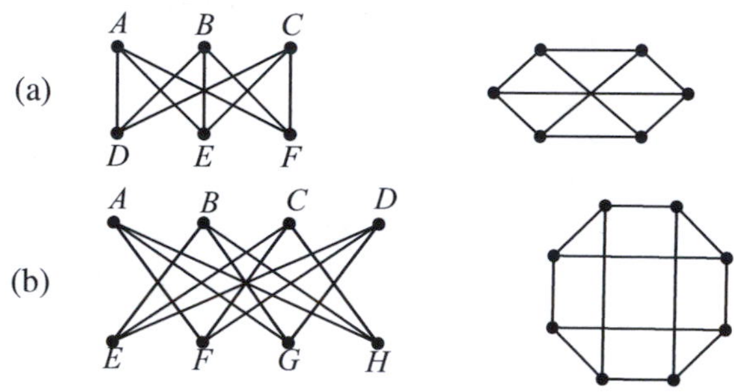

Matrices Associated with a Graph

The essential information about a graph can be recorded in a matrix consisting of 0's and 1's in two different ways. The *adjacency matrix* of the graph indicates which vertices are adjacent to each other, while the *incidence matrix* indicates which edges occur at each vertex. For example, the matrices associated with the graph at the left are shown here.

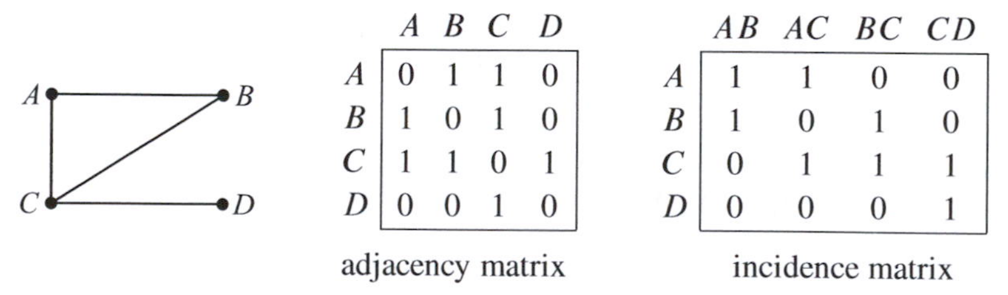

	A	B	C	D
A	0	1	1	0
B	1	0	1	0
C	1	1	0	1
D	0	0	1	0

adjacency matrix

	AB	AC	BC	CD
A	1	1	0	0
B	1	0	1	0
C	0	1	1	1
D	0	0	0	1

incidence matrix

A24 Find adjacency and incidence matrices associated with this graph.

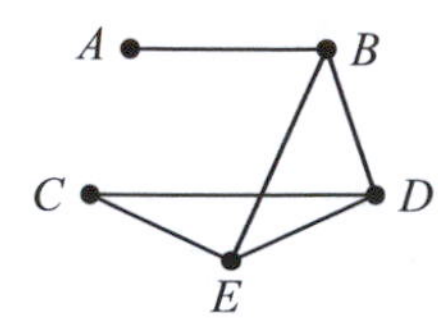

A25 What is the significance of the number of 1's that occur in each row or column of the adjacency matrix of a given graph? In each row of the incidence matrix? In each column of the incidence matrix?

A26 Use the incidence matrix of a graph to restate the proof of the Degree Theorem. (Hint: Count the number of 1's in the matrix in two different ways.)

A27 Suppose that **G** is a graph and **H** is a subgraph of **G**. Using the symbols $\delta_{\mathbf{G}}$, $\delta_{\mathbf{H}}$, $\Delta_{\mathbf{G}}$ and $\Delta_{\mathbf{H}}$ in the obvious way, consider the two statements

$$(1)\quad \delta_{\mathbf{G}} \geq \delta_{\mathbf{H}}, \qquad \text{and} \qquad (2)\quad \Delta_{\mathbf{G}} \geq \Delta_{\mathbf{H}}.$$

(a) Only one of these statements is true for all graphs. Which is it?
(b) Find a counterexample showing that the other statement is not always true.

A28 Describe all possible 2-regular connected graphs.

A29 (a) Find a 2-regular graph that has seven vertices and two components.
(b) Describe all possible 2-regular graphs having any number of components.

A30 (a) If a 5-regular graph has 100 vertices, then how many edges does it contain?
(b) If a 5-regular graph has 100 edges, then how many vertices does it contain?

A31 (a) Show that if a graph is d-regular and has n vertices and e edges, then $nd = 2e$.
(b) Use the equation established in (a) to find the number of edges in the complete graph $\mathbf{K}_{10}$.
(c) Find a general formula for the number of edges in the complete graph $\mathbf{K}_n$.

A32 Suppose that a graph contains eight vertices and 21 edges.
(a) Use the Degree Theorem to find the average of all of the degrees in this graph.
(b) Explain how you know that this graph must contain at least one vertex that has degree less than or equal to 5 and at least one vertex that has degree greater than or equal to 6.

A33 Use the result of problem A16 to show that there exists a graph having the degree sequence $(5, 5, 4, 4, 4, 3, 3)$ without actually drawing such a graph. (Hint: Think about complementary graphs.)

A34 Use complementary graphs to show that there exists a graph having the degree sequence $(5, 5, 5, 5, 4, 4, 4)$.

The Degree Sequence Algorithm

The following procedure can be used to determine whether a given sequence of numbers is the degree sequence of some graph.

- Start with the terms of the sequence in decreasing order.
- Remove the largest term m, and reduce the next m terms each by 1.
- Rearrange the new sequence in decreasing order (if necessary).

The significance of this procedure is that the original sequence is the degree sequence of some graph if and only if the new sequence is. The process can then be repeated until a conclusion is reached.

As an illustration of how the Degree Sequence Algorithm works, we apply it to the sequence in problem A34. Comments on each step appear at the right.

5555444	$m = 5$, so remove the first term and reduce the next 5 terms
444334	rearrange to decreasing order
444433	$m = 4$, so remove the first term and reduce the next 4 terms
33323	rearrange to decreasing order
33332	$m = 3$, so remove the first term and reduce the next 3 terms
2222	

At this point it becomes obvious that we have the degree sequence of a graph: $(2, 2, 2, 2)$ is the degree sequence of the cycle graph $\mathbf{C}_4$. Therefore we conclude that the original sequence is also the degree sequence of some graph. In fact, the Degree Sequence Algorithm provides a procedure for constructing such a graph. The next problem illustrates this.

A35 (a) Starting with $\mathbf{C}_4$, add a vertex to produce a graph having the degree sequence $(3, 3, 3, 3, 2)$.
(b) Next add a vertex to produce a graph having the degree sequence $(4, 4, 4, 4, 3, 3)$.
(c) Finally add one more vertex to produce a graph having the degree sequence $(5, 5, 5, 5, 4, 4, 4)$.

In other words, whenever the Degree Sequence Algorithm leads to the degree sequence of a known graph, we can construct a graph for the original sequence by working backwards.

A36 Use the procedure illustrated by problem A35 to construct a graph having the degree sequence in problem A33.

A37 Use the Degree Sequence Algorithm to show that there is no graph having any of the following degree sequences:
(a) (5, 5, 5, 5, 2, 2, 2).
(b) (7, 7, 6, 5, 5, 3, 3, 2)
(c) (8, 8, 8, 7, 6, 4, 4, 3, 2, 2).

A38 (a) Without using the Degree Sequence Algorithm, explain why it is impossible for a graph with n vertices to have the degree sequence $(n-1, n-2, \ldots, 1, 0)$ if n is greater than or equal to 2.
(b) Prove that in any graph having two or more vertices, there must exist at least two vertices having the same degree. (Hint: If all the degrees were different, then what would the degree sequence be?)

A39 (a) Suppose that an edge is added to a graph, joining two existing vertices that were previously nonadjacent. What can happen to the number of components in this graph as a result of adding the edge? Give examples to illustrate each possibility.
(b) Answer the same question if an edge is removed from a graph but its endpoints remain.

A40 Suppose that A and B are two vertices in a given graph and suppose that there is some path in the graph starting at A and ending at B. Does there necessarily exist a simple path from A to B in this graph? How do you know this? (Suggestion: See problem A3.)

Definitions The *distance* between two vertices in the same component of a graph is the length of a shortest path joining these vertices. If **G** is a connected graph, then the *diameter* of **G** is the maximum of all of the distances between vertices of **G**.

A41 What, if anything, does this definition have in common with the definition of the diameter of a circle?

A42 Which graphs have diameter equal to 1?

A43 Find the diameter of each graph below.

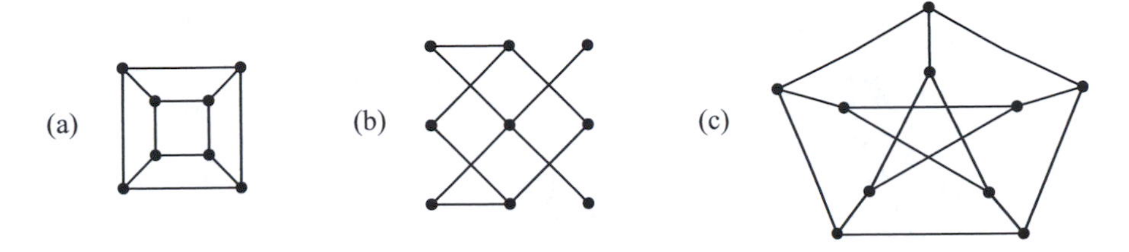

A44 Prove that if $\mathbf{G}_1$ and $\mathbf{G}_2$ are complementary graphs, then at least one of them must be connected. (Hint: Show that if $\mathbf{G}_1$ is disconnected, then the diameter of $\mathbf{G}_2$ is at most 2.)

A45 Suppose that a graph contains 11 vertices, and that each vertex has degree 5 or greater. Prove that the graph must be connected.

A46 Show that in a graph with minimum degree δ, each component must contain at least $\delta + 1$ vertices.

A47 Suppose that a graph has 15 vertices and the minimum degree is 3. Prove that the number of components is at most 3.

A48 Show that in a graph with n vertices, if $\delta + \Delta$ is greater than or equal to $n - 1$, then the graph must be connected. (Hint: First show that there must be some component that contains at least $\Delta + 1$ vertices. Then show that if there were any other component, these two components together would contain more than n vertices.)

In the problems below, let P be a longest simple path in a graph **G**, and let λ represent the length of P. (P contains λ edges and $\lambda + 1$ vertices.)

A49 Show that each endpoint of P has degree less than or equal to λ. (Hint: Where can edges go to from these vertices?)

A50 Prove that in any graph with minimum degree δ, there exists a simple path of length greater than or equal to δ. (Hint: This is equivalent to showing that λ is greater than or equal to δ.)

A51 Assume that only one vertex in **G** has degree δ. Then how can the result in the preceding problem be improved?

Isomorphic Graphs

Consider these two labeled graphs:

Although they are not equivalent, both correspond to the same unlabeled graph:

Therefore, in some sense, the two labeled graphs are closely related. To put it another way, it is possible to re-label these two graphs in such a way that they become equivalent. This leads to the concept of *isomorphic graphs*: Two labeled graphs are *isomorphic* to each other if they can be re-labeled in such a way that they become equivalent graphs. (Obviously it would be enough to just re-label one of the two graphs. Also, if two graphs are already equivalent, then the "re-labeling" doesn't necessarily have to change anything. So equivalent graphs are considered to be isomorphic to each other.)

B1 Are these graphs isomorphic?

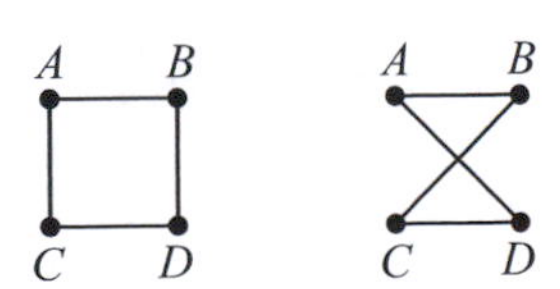

What about these?

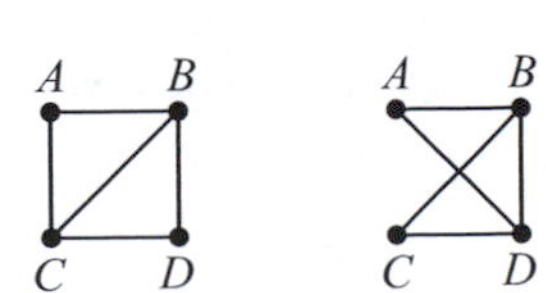

So far we have defined isomorphic labeled graphs. Two unlabeled graphs are considered isomorphic if it is possible to label their vertices in such a way that they become equivalent graphs.

B2 Show that these two graphs are isomorphic.

B3 Classify each statement as either true or false. In each assume that the graphs are labeled.

(a) All complete graphs are equivalent.

(b) All complete graphs with the same number of vertices are equivalent.

(c) All complete graphs with the same number of vertices are isomorphic.

(d) All cycle graphs with the same number of vertices are equivalent.

(e) All cycle graphs with the same number of vertices are isomorphic.

If two graphs are isomorphic, then they have the same number of vertices, the same number of edges, the same maximum and minimum degrees, the same degree sequence, the same number of components, the same diameter (defined before problem A49), and the same length of a longest simple path. So if two graphs differ in any of these respects, then they are not isomorphic. However, having all of these values in common does not necessarily indicate that two graphs are isomorphic.

B4 Show that these two graphs are not isomorphic.

When listing unlabeled graphs, we usually want to avoid including any graphs that are isomorphic to each other. For example, below are all of the unlabeled graphs having at most three vertices, including no isomorphic repetitions:

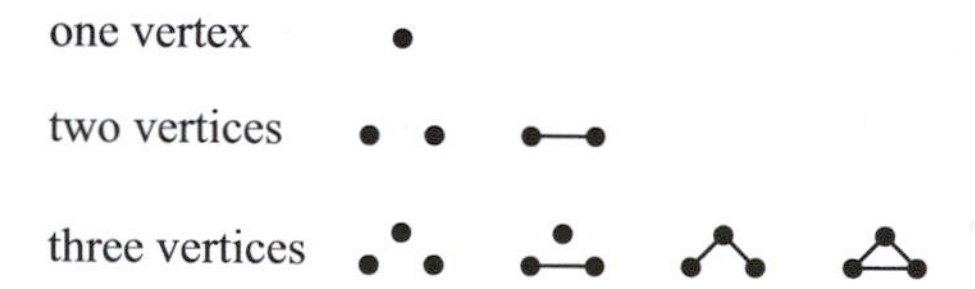

When we reach four vertices, things get more interesting.

B5 List all of the unlabeled graphs with four vertices, including no isomorphic repetitions. There are 11 of them.

B6 The graphs in problem B5 can be grouped into complementary pairs, except for one which is isomorphic to its own complement. Indicate which graphs are grouped together.

B7 Of the 11 graphs found in problem B5, count the number of these having each number of edges, 0 through 6. What do you notice? Explain why this happens.

If two graphs are isomorphic and one of them contains a cycle of a particular length, then the same must be true of the other graph.

B8 Show that these two graphs are not isomorphic by considering lengths of cycles contained within each.

In some cases it is helpful to consider complementary graphs when trying to determine whether graphs are isomorphic. The relevant fact is that two graphs are isomorphic to each other if and only if their complements are isomorphic to each other.

B9 Use complementary graphs to show that these two graphs are isomorphic and find a labeling of the vertices that makes them equivalent.

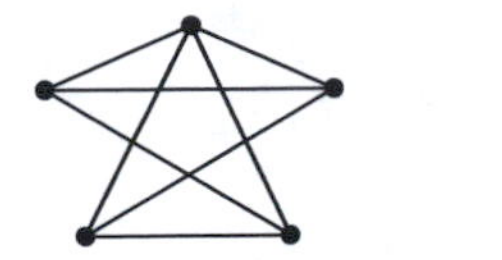

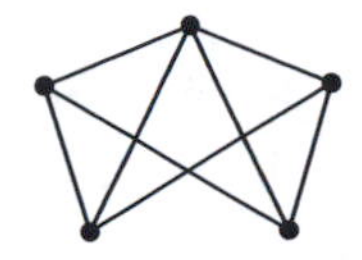

B10 Use complementary graphs to determine which of these six graphs are isomorphic to each other.

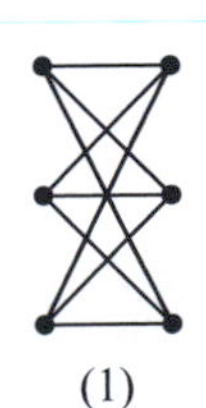

(1)

(2)

(3)

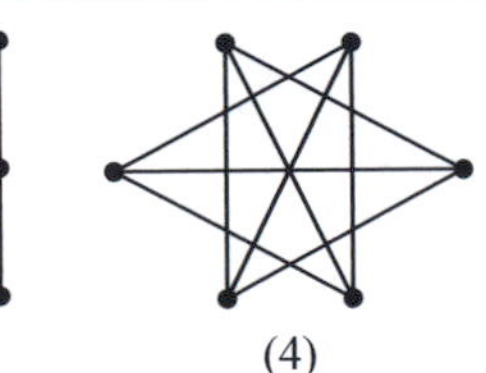

(4)

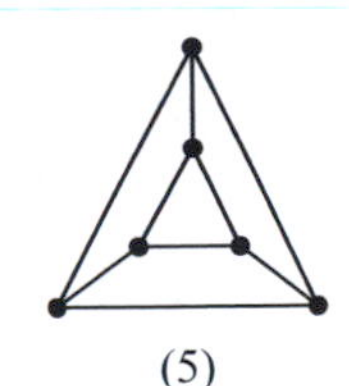

(5)

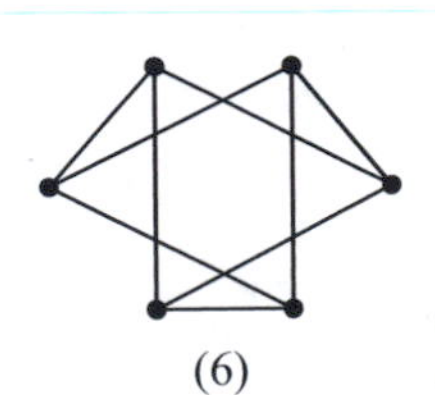

(6)

More Problems

B11 List all of the unlabeled graphs with five vertices and at most five edges, including no isomorphic repetitions.

B12 Use your answer from problem **B11** to determine the total number of different (non-isomorphic) unlabeled graphs with five vertices.

B13 List all of the 2-regular unlabeled graphs with 11 vertices, including no isomorphic repetitions.

B14 Show that all 6-regular graphs with eight vertices are isomorphic to each other. Draw one of these graphs.

B15 Show that there are exactly two different (non-isomorphic) 4-regular unlabeled graphs with seven vertices. Draw one of each.

B16 Show that these two graphs are isomorphic.

 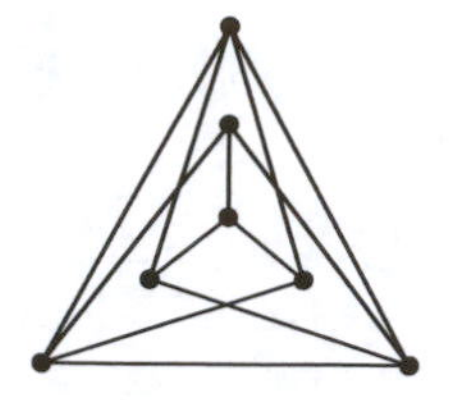

B17 Show that these two graphs are not isomorphic.

 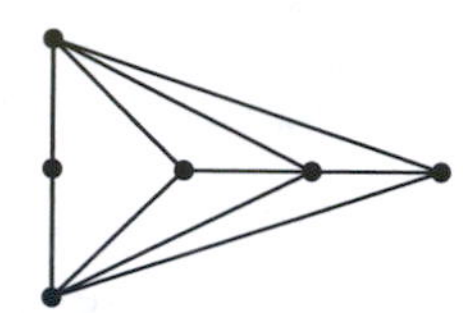

B18 Which of these graphs, if any, are isomorphic to each other?

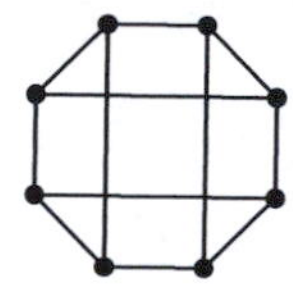  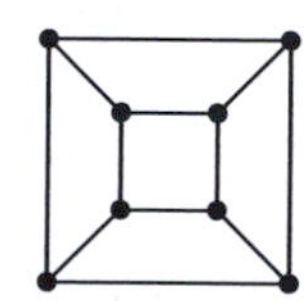

B19 In this set of seven graphs

(1)

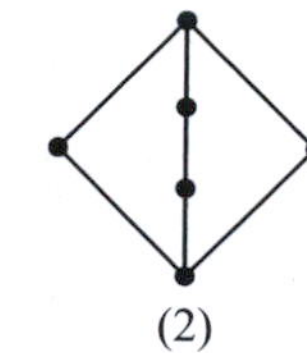
(2)

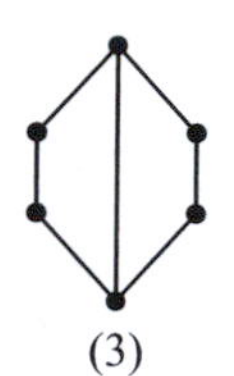
(3)

(4)

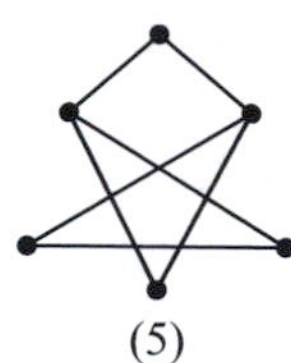
(5)

(6)

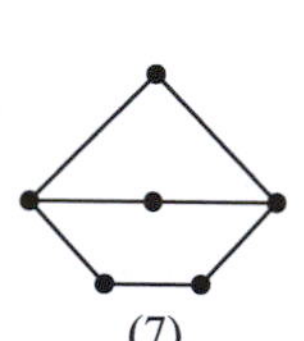
(7)

(a) Find four that are all isomorphic to each other.

(b) Find three that are mutually non-isomorphic.

B20 Two graphs that look as different as these couldn't possibly be isomorphic. Or are they?

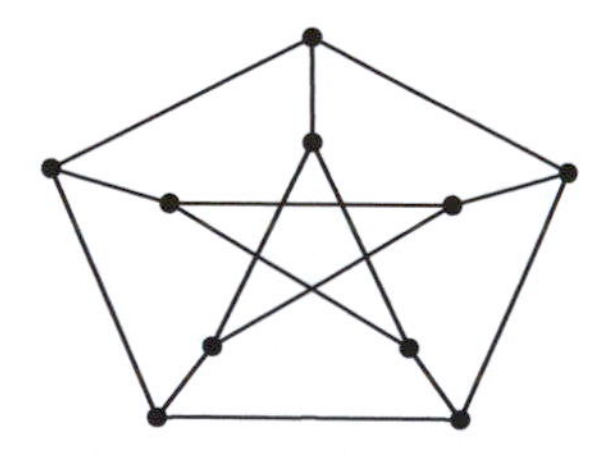 

B21 Which of these graphs, if any, are isomorphic to each other?

 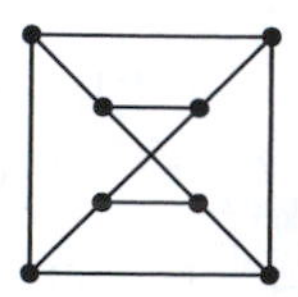 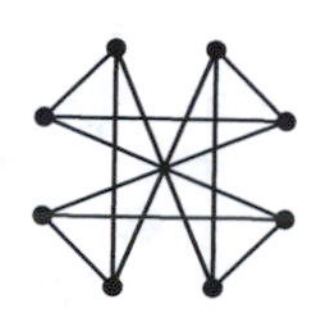

C

Bipartite Graphs

A *bipartite graph* is a graph whose vertices can be separated into two sets **X** and **Y** in such a way that every edge in the graph has one endpoint in each set.

Examples

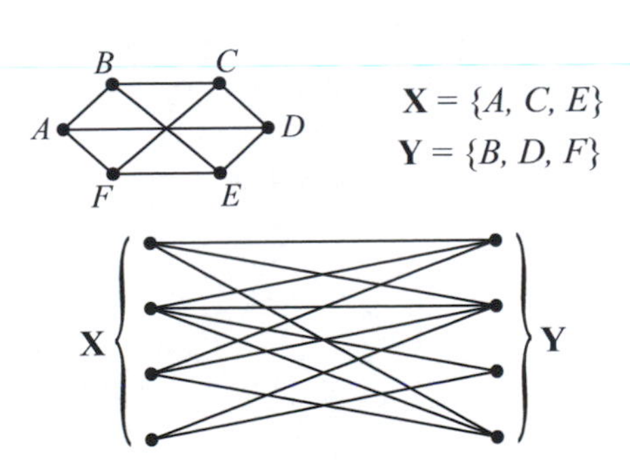

Another way to say this is that a graph is bipartite if it is *2-colorable*: This means that each vertex can be assigned one of two colors (for example, red or blue) in such a way that each edge in the graph has one endpoint of each color.

C1 Which of these are bipartite graphs?

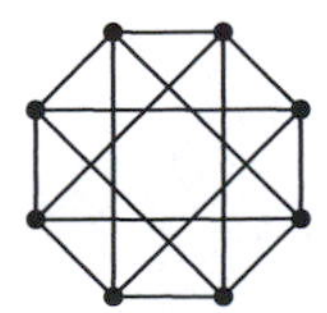 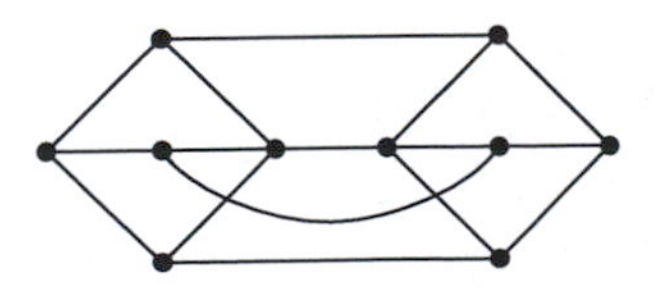 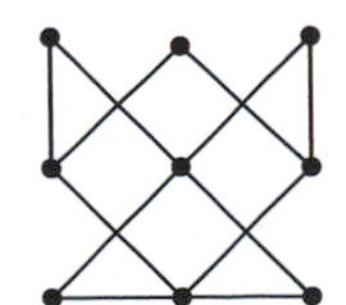 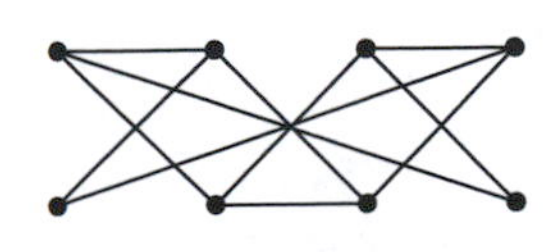

A few small details need to be clarified here. First, a graph containing only one vertex qualifies as bipartite: It is not necessary for both colors to actually appear. Equivalently, one set, **X** or **Y**, can be empty. Also, any graph that has no edges is bipartite. This can be seen more easily if we reword the definition of a bipartite graph to say that no edge has both of its endpoints in the same set, **X** or **Y**. In terms of coloring, no edge has two endpoints of the same color. Clearly a graph with no edges satisfies this condition no matter how the vertices are colored.

C2 For which values of n is a cycle graph C_n bipartite?

C3 Suppose that **G** is a bipartite graph and **H** is a subgraph of **G**. Explain how you know that **H** is also bipartite.

Complete Bipartite Graphs

There are special bipartite graphs having the property that every vertex in **X** is adjacent to every vertex in **Y**. These are called *complete bipartite graphs* and are denoted by the symbol $\mathbf{K}_{m,n}$, where m and n are the numbers of vertices in **X** and **Y**.

Examples

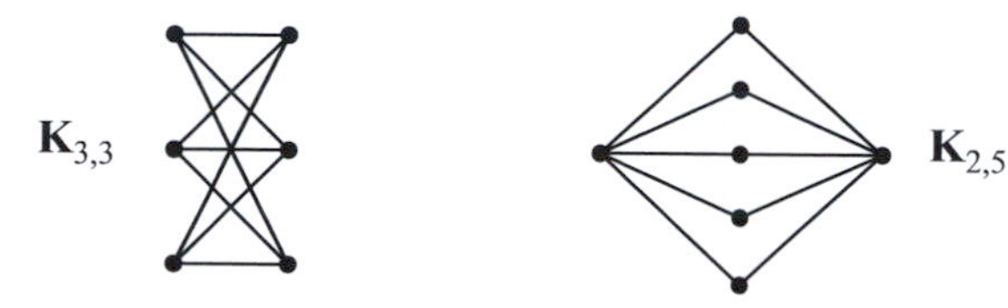

C4 How many edges does $\mathbf{K}_{m,n}$ contain? What are the degrees of the vertices?

C5 Color the vertices of this bipartite graph using red and blue so that each edge has one endpoint of each color, and compare three things: the sum of the degrees of all the red vertices, the sum of the degrees of all the blue vertices, and the number of edges in the graph. Explain why all of these values are equal.

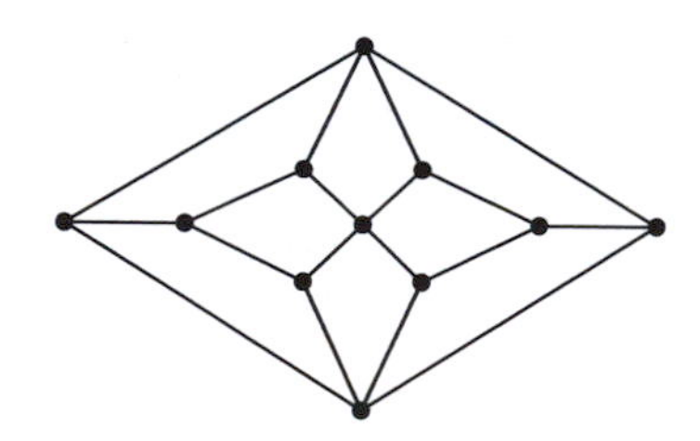

C6 Restate the result of the preceding problem in terms of the two sets **X** and **Y** in the definition of a bipartite graph.

C7 Show that a 4-regular graph having 15 vertices cannot be bipartite.

C8 Find a bipartite graph having the degree sequence $(4, 3, 3, 3, 3, 3, 3, 2, 2)$.

C9 Show that there is no bipartite graph having the degree sequence $(6, 6, 6, 4, 4, 4, 4, 4, 4, 4, 4, 4, 4)$.

Bipartite Graphs and Matrices

A bipartite graph can be represented by a matrix in a way that is more efficient than the adjacency and incidence matrices introduced in chapter A. The example below illustrates how this is done.

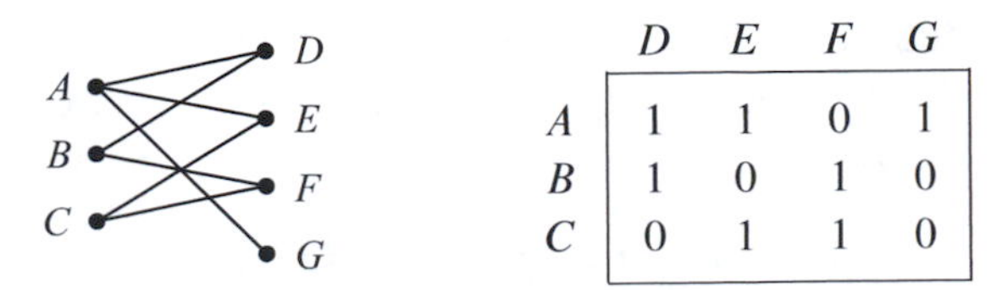

	D	E	F	G
A	1	1	0	1
B	1	0	1	0
C	0	1	1	0

C10 Represent this bipartite graph by a matrix. First label the vertices.

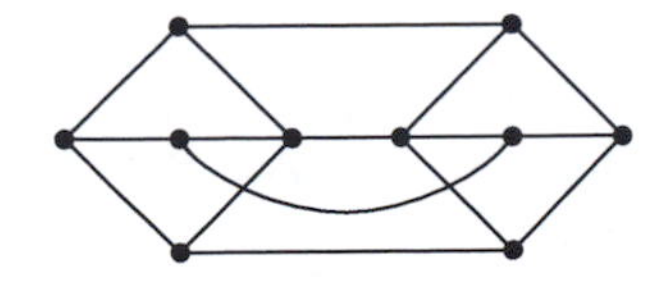

C11 In the correspondence between a bipartite graph and a matrix, each edge of the graph corresponds to a 1 in some position of the matrix. Suppose that two edges have no common endpoint. How is this property reflected in the positions of the corresponding 1's?

C12 Find a set of five edges in this graph, no two of which have a common endpoint.

(a) by trial and error, working only with the graph.

(b) by looking for a set of appropriately positioned 1's in the corresponding matrix.

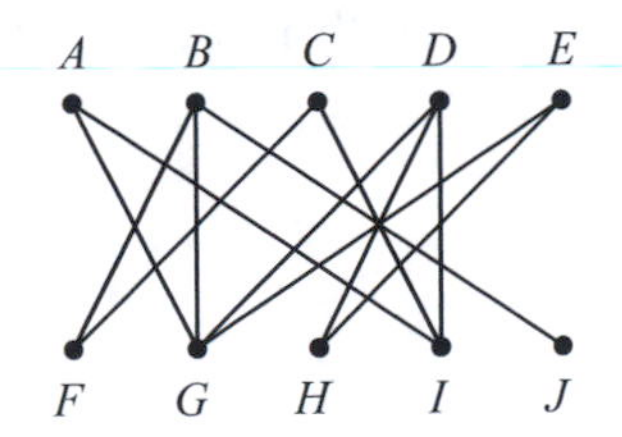

Cycles in a Bipartite Graph

C13 Show that if a graph contains at least one odd cycle, then the graph is not bipartite.

The statement above provides an easy way to prove that a given graph is not bipartite: Just find any odd cycle in the graph. To prove that a graph *is* bipartite, the easiest way is usually to color the vertices appropriately using two colors. However, it is interesting that the converse of the statement in problem C13 is also true: If a graph is not bipartite, then it contains at least one odd cycle. Or equivalently, we can say it this way:

Cycle Theorem for Bipartite Graphs

A graph is bipartite if and only if it contains no odd cycles.

We already know that if a graph is bipartite then it contains no odd cycles. This is logically equivalent to the statement in problem C13. The harder part is to show that if a graph contains no odd cycles, then it is bipartite.

To explain why this is true, it will be helpful to introduce another concept, that of an odd closed path. Recall that a closed path starts and ends at the same vertex. An *odd closed path* is a closed path of odd length, which means that it contains an odd number of edges.

C14 Let $A, B, \ldots, J$ be 10 distinct vertices of a graph such that $ABCDEFCGBHCIJA$ is a path. This is a closed path of length 13. Find two shorter odd closed paths. Is either one of these a cycle?

C15 Let $V_1 V_2 \ldots V_n V_1$ be an odd closed path in a graph **G**. Show that either $V_1 V_2 \ldots V_n V_1$ is a cycle or else **G** contains a shorter odd closed path. (Suggestion. Separate two cases: Either $V_1, V_2, \ldots, V_n$ are distinct vertices or else they contain some repetition.)

C16 Prove that if a graph contains an odd closed path, then it contains an odd cycle. (Hint: Consider a shortest odd closed path.)

C17 Show that if each component of a graph is bipartite, then the entire graph is bipartite.

Proof of the Cycle Theorem

Let **G** be a graph that contains no odd cycles. In the following problems we will prove that **G** is bipartite. In view of problem C17, it will be enough to show that each component of **G** is bipartite.

Starting at any vertex A in any component of **G**, assign the color red to A and proceed to color vertices along simple paths from A, alternating between red and blue. In that way every vertex in the component is eventually reached and colored.

C18 Try it for this graph.

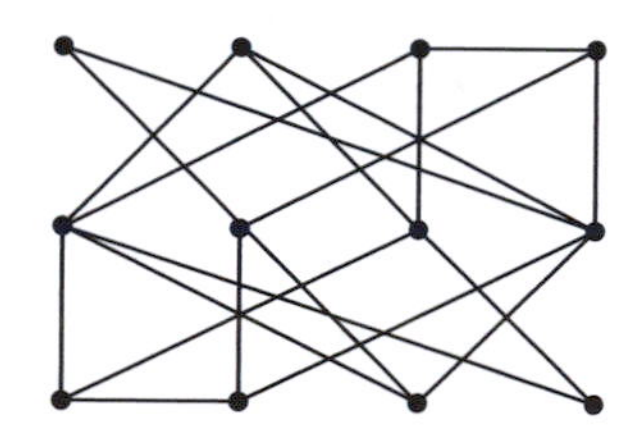

In the procedure described above, the color assigned to a vertex V depends on the length of the path followed from A to V. A vertex reached by a path of odd length is colored blue, while a vertex reached by a path of even length is colored red. The next problem shows that there is only one possibility for the coloring of a vertex reached from A.

C19 Let A and V be vertices in a graph that contains no odd cycles and suppose that there are two paths from A to V, one of odd length and one of even length. We will show that this situation is impossible by deriving a contradiction.

(a) Show that the graph contains an odd closed path.

(b) Find the contradiction.

C20 Complete the proof that **G** is bipartite. Show that with the vertices of some component colored as described above, no edge has the same color assigned to both of its endpoints.

More Problems

C21 Which of these are bipartite graphs?

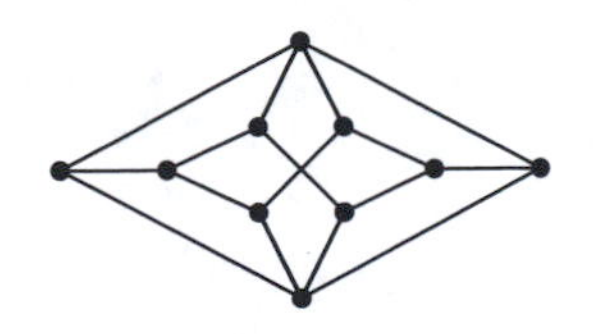

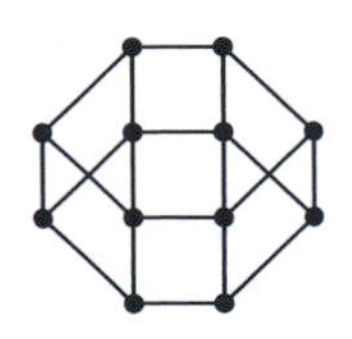

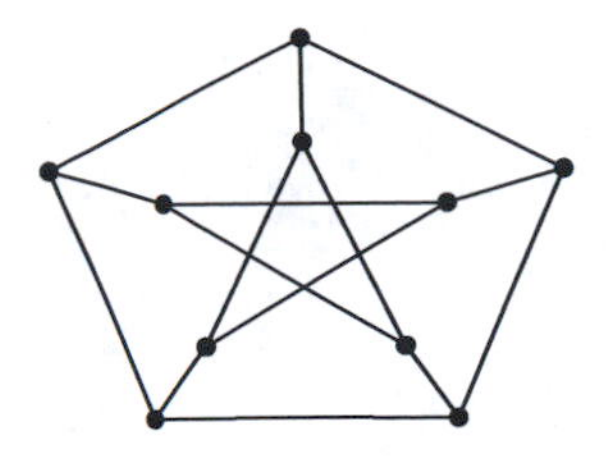

C22 Suppose that two graphs are isomorphic and one of them is bipartite. Then must the other graph necessarily also be bipartite? Explain.

C23 Show that a graph that has 17 vertices and 73 edges cannot be bipartite.

C24 Suppose that in a bipartite graph, all of the vertices except possibly one of them have the same degree d, and the remaining vertex has an unknown degree x. Prove that x must be a multiple of d (0, d, $2d$, $3d$, etc.)

C25 Find all bipartite graphs containing exactly five vertices and at least one cycle. (Suggestion: Think of what cycles can exist in the graph.)

C26 For each sequence below, decide whether there exists a bipartite graph with these degrees. If so, find one. If not, explain why.

(a) (6, 6, 4, 4, 4, 4, 4, 4, 2, 2, 2)
(b) (4, 4, 4, 4, 4, 3, 3, 2, 2)
(c) (5, 5, 5, 5, 5, 5, 4, 4, 4)
(d) (5, 4, 4, 3, 3, 2, 2, 2, 1)
(e) (5, 5, 4, 3, 3, 3, 3, 3, 3)
(f) (5, 5, 5, 5, 4, 4, 4, 3, 3)

C27 Suppose that **G** is a bipartite graph with a particular division of the vertices into two sets **X** and **Y**, as in the definition of a bipartite graph. Make up a reasonable definition for the *bipartite complement* of **G**, another bipartite graph with the same vertices.

C28 Use the idea of the bipartite complement of a graph to solve problem C8 by replacing the degrees with smaller numbers. Can you do the same for any parts of problem C26?

C29 Use a matrix to find a set of five edges in this graph, no two of which have a common endpoint.

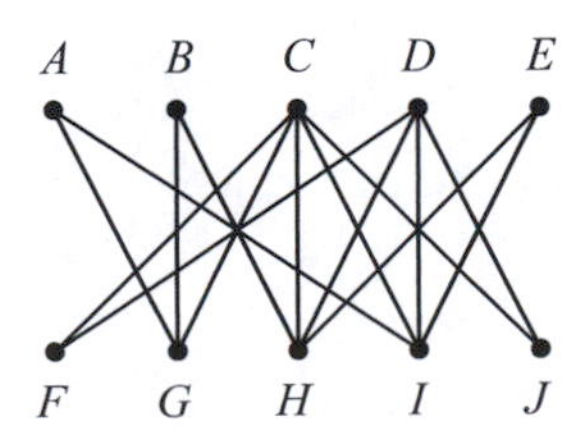

Trees and Forests

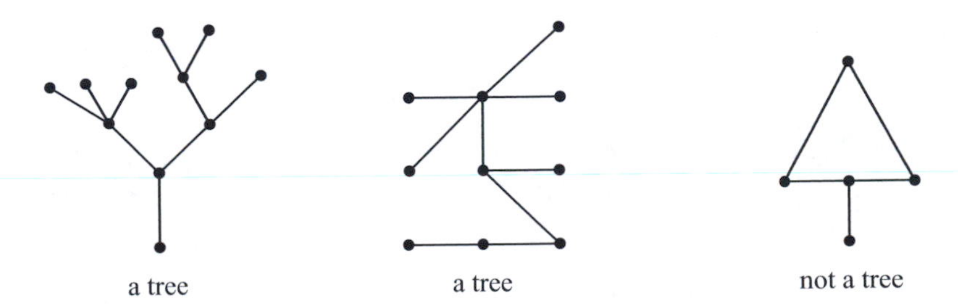

Definitions A *tree* is a graph that is connected and contains no cycles. A *forest* is a graph in which each component is a tree. (This allows the possibility that the graph consists of a single tree.) An equivalent definition of a forest would be a graph that contains no cycles.

Below is a list of all of the unlabeled trees containing up to five vertices, avoiding isomorphic repetitions.

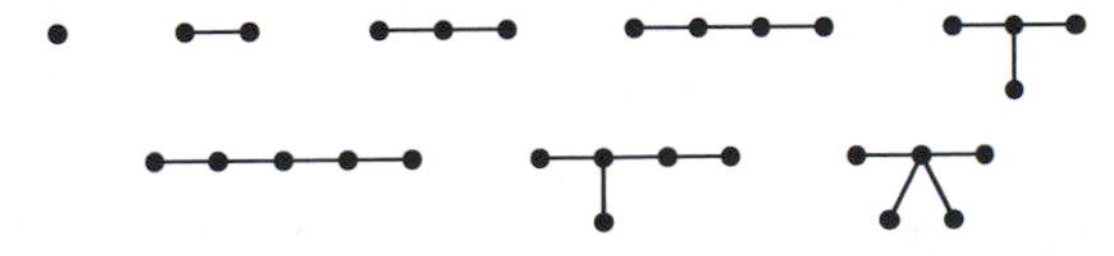

D1 List all of the unlabeled trees with six vertices, avoiding isomorphic repetitions.

D2 What relationship seems to exist between the number of vertices and the number of edges in a tree?

D3 Find all of the unlabeled forests with five vertices and two or more components. As usual, avoid isomorphic repetitions.

D4 From the examples found in the previous problem, what relationship seems to exist between the number of vertices, number of edges, and number of components in a forest? Use the notation v for the number of vertices, e for the number of edges, and k for the number of components.

Pruning a Tree

In order to prove theorems about trees, it is helpful to introduce the process of *pruning* a tree: This means removing a vertex of degree 1 from the tree along with the edge that occurs at that vertex. However, the other endpoint of that edge remains in the tree. This step can then be repeated any number of times.

Example of pruning a tree

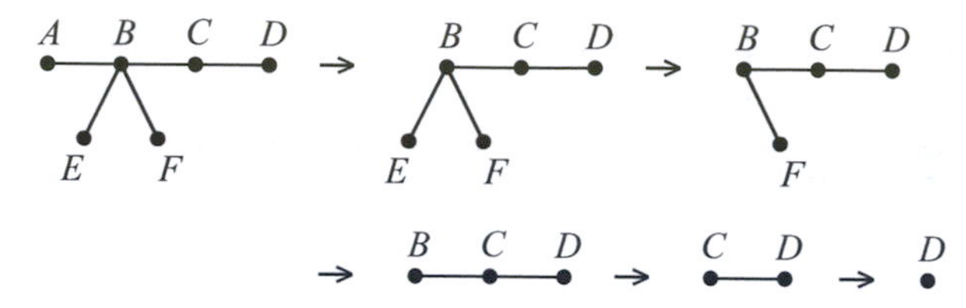

D5 Pruning Lemma Prove that on each step of the pruning process, when one vertex and one edge are removed from a tree, the remaining graph is still a tree. Use the definition of a tree.

We need another fact: that every tree with more than one vertex can be pruned. This fact depends on the existence of a vertex of degree 1 in the tree.

D6 Lemma on Vertices of Degree 1 Prove that every tree with two or more vertices has at least two vertices of degree 1. (Suggestion: Consider a longest simple path in the tree. What must be true about its endpoints, and why?)

Now we can prove something that we observed previously:

Tree Theorem 1 *Every tree with v vertices has exactly $v - 1$ edges.*

This is equivalent to saying that the difference $v - e$ is always 1 for a tree, where v is the number of vertices and e is the number of edges.

The proof of Tree Theorem 1 is accomplished by answering the following two questions.

D7 What happens to the difference $v - e$ on each step of the pruning process?

D8 What is the value of $v - e$ at the end of the pruning pocess, when the tree has been reduced down to a single vertex? What does that mean about the original value of $v - e$?

Corollary *In a forest with v vertices and k components, the number of edges is $v - k$.*

D9 Show how the statement above follows from Tree Theorem 1. (Suggestion: Consider the difference $v - e$ for each component tree.)

D10 Suppose that a graph has 20 vertices, 15 edges, and contains no cycles. Then how many components does the graph contain?

D11 Suppose that a graph contains an equal number of vertices and edges. Show that the graph must contain at least one cycle. Is the graph necessarily a cycle graph?

Tree Theorem 2 *In any tree, there is exactly one simple path from any vertex to any other vertex.*

This statement may appear to be obvious, but it requires some justification. Let A and B represent two vertices in a tree $\mathbf{T}$. We have to show that a simple path from A to B exists in $\mathbf{T}$ and that such a path is unique.

D12 (existence) Indicate how you know that $\mathbf{T}$ contains a simple path from A to B.

D13 (uniqueness) Try to explain why there could not be two different simple paths from A to B in $\mathbf{T}$. (This may not be quite as easy as it sounds. Take into account that two different paths might intersect in vertices other than just A and B.)

A uniqueness proof that avoids the complication of intersecting paths is given in problem D35.

Directed Trees

A *directed tree* is a tree in which each edge includes a direction from one endpoint to the other. In other words, it is a directed graph that becomes a tree in our usual sense when all of the directions are removed.

One vertex B in a directed tree is *reachable* from another vertex A if the tree contains a path from A to B that follows all of its edges in the specified directions. For example, in the directed tree below, vertex A is reachable from E but not reachable from C.

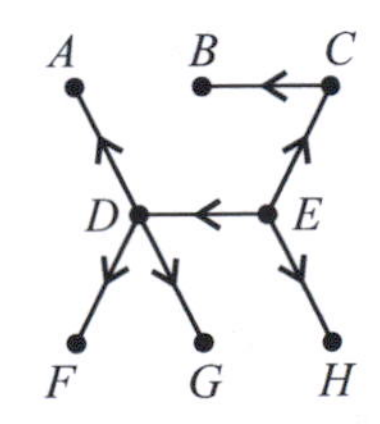

If a directed tree contains a vertex R from which all other vertices are reachable, then R is called a *root* for that tree. In the example above, E is a root.

A directed tree may or may not contain a root.

D14 Find an example of a directed tree that has no root and observe the number of incoming edges that occur at each vertex. Then do the same for the example above in which vertex E is a root. What do you notice?

Problem D14 suggests that the existence of a root in a directed tree is related to the number of incoming edges (the *indegree*) at each vertex of the tree.

D15 Let $\mathbf{T}$ be a directed tree that contains a root R. Prove the following statements.

(a) At each vertex of $\mathbf{T}$ other than R, the indegree is at least 1.

(b) The indegree at R is 0, and at each vertex of $\mathbf{T}$ other than R, the indegree is exactly 1. (Hint for (b): How many edges exist in $\mathbf{T}$, and how is this number related to the sum of all the indegrees?)

Problem D15 shows that if a directed tree contains a root, then at all but one vertex of the tree the indegree must be 1, and at the remaining vertex the indegree is 0. Moreover that remaining vertex must be the root. However, what about the converse of this statement? If a directed tree has indegrees as described above (one 0 and all the rest 1s), does the tree necessarily contain a root? Or more specifically, are all vertices of the tree reachable from the vertex that has indegree 0? Although this is not obvious, it is true.

The Root Theorem *A directed tree contains a root if and only if all but one vertex has indegree 1 and the remaining vertex has indegree 0. If this condition is satisfied, then the vertex having indegree 0 is the root.*

Proof One direction of this theorem has already been established in problem D15. For the other direction (the existence of a root whenever the indegree condition is satisfied), let R be the vertex having indegree 0 and let V be any other vertex in the tree. We claim that V is reachable from R. Temporarily ignoring the directions of the edges, we know that the tree contains an undirected path **P** from R to V. It remains to show that when the directions are considered, **P** is a directed path.

D16 Use the indegree condition to prove that **P** is a directed path from R to V.

Spanning Trees

In any connected graph **G**, a *spanning tree* is a subgraph of **G** having the following two properties:

1. The subgraph is a tree, and
2. The subgraph contains every vertex of **G**. For example, in this graph

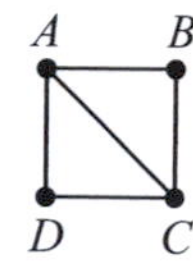

one of the spanning trees is

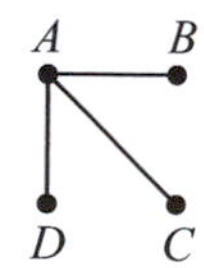

D17 There are seven more spanning trees in the given graph. Find all of them.

Because of the labels on the vertices, the trees found in problem D17 are all different (nonequivalent) graphs, although many of these are isomorphic to each other.

Counting Spanning Trees

Suppose that we want to know how many spanning trees exist in a given labeled graph. This question is of interest, for example, in the theory of electrical networks. There is a simple formula for the number of such trees in the case where the given graph is complete. In other cases it becomes more complicated.

Cayley's Formula In a labeled complete graph with n vertices, the number of spanning trees is n^{n-2}.

D18 Verify by counting trees that this is correct for $n = 1, 2, 3$, and 4.

Recall that in problem **D17** we found eight spanning trees in this graph **G**.

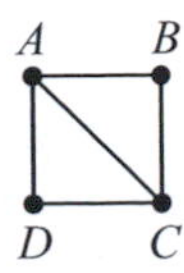

There is an interesting way of looking at that result in connection with Cayley's Formula. First, notice that the spanning trees in **G** are spanning trees in $\mathbf{K}_4$ that do not contain the edge BD. Each spanning tree in $\mathbf{K}_4$ contains 3 out of the 6 edges of $\mathbf{K}_4$. This suggests that each edge of $\mathbf{K}_4$ (in particular, BD) has a 50% chance of being in any given spanning tree. To put it another way, we would expect that exactly half of the 16 spanning trees of $\mathbf{K}_4$ contain the edge BD. The other half do not, so they are spanning trees in **G**.

D19 Use the same reasoning to determine the number of spanning trees in this graph.

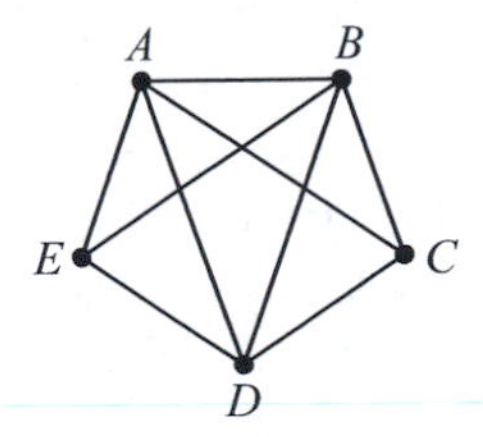

Codewords for Trees: Prufer's Method

In order to establish Cayley's Formula, we assign a certain sequence, or *codeword*, to each spanning tree in $\mathbf{K}_n$. There will be an easy way to count the codewords, and because each tree will be given a unique codeword, there will be an equal number of spanning trees. To illustrate this, let $n = 5$.

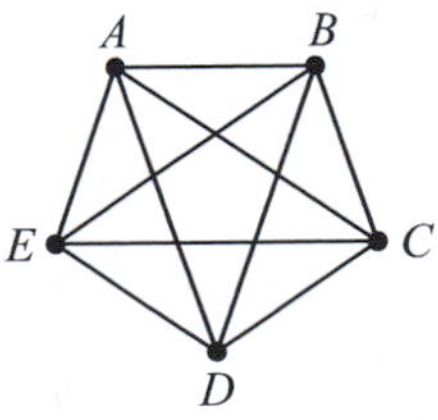

For each spanning tree in $\mathbf{K}_5$, we first assign an 8-letter codeword by pruning the tree and recording the endpoints of each edge as the edge is removed. However, the pruning takes place in a particular order. First, we assign a particular ordering to the vertices. In this example we will use the usual alphabetical order A, B, C, D, E. Then on each step, the pruning takes place at the vertex of degree 1 that comes earliest in the ordering. That vertex and the edge that contains it are removed.

Example

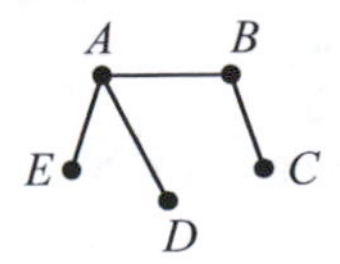

On the first step, vertices C, D, and E each have degree 1. C comes earliest in the ordering, so the first edge to be pruned is CB.

D20 Continue this process, obtaining the codeword $CBBADAAE$.

D21 Obtain the 8-letter codeword that corresponds to this tree.

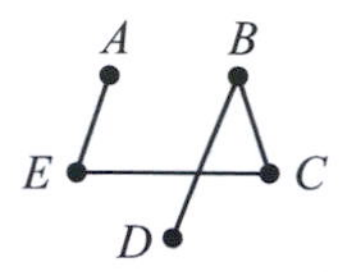

Going the opposite way, from the 8-letter codeword to the tree, is easy since the edges can be read directly from the codeword: The first two letters determine one edge, the next two determine another, etc. Now, however, comes the tricky part. Not every 8-letter codeword corresponds to a tree, and in fact the key information is contained in just three of the positions: the 2nd, 4th and 6th. These three letters form what we call the *short codeword* corresponding to the tree. (The 8-letter codeword is called the *long codeword*.) In problem D20 the short codeword is BAA. In problem D21 it is EBC.

D22 Find the long and short codewords corresponding to this tree.

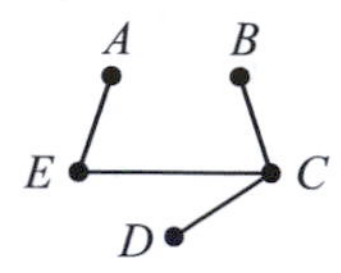

D23 What relation exists between the degree of a vertex in a tree and the number of times that vertex occurs in the long codeword corresponding to the tree? In the corresponding short codeword?

D24 By trial and error, find a spanning tree in $\mathbf{K}_5$ that corresponds to the short codeword BBD.

There is a systematic procedure for solving problems like D24 (see problem D45), and it leads to the conclusion that there is a one-to-one correspondence between the spanning trees in $\mathbf{K}_5$ and all possible 3-letter codewords using letters from the set $\{A, B, C, D, E\}$. There are 5^3, or 125, such codewords, and therefore an equal number of spanning trees in $\mathbf{K}_5$. More generally, in $\mathbf{K}_n$, each spanning tree corresponds to a short codeword containing $n-2$ symbols chosen from the set of n vertices of $\mathbf{K}_n$. The number of these codewords is n^{n-2}, and therefore the number of spanning trees in $\mathbf{K}_n$ is the same.

More Problems

D25 Find all unlabeled forests that have six vertices and two or more components. Avoid isomorphic repetitions.

D26 Find all unlabeled trees that have seven vertices. Avoid isomorphic repetitions. (Suggestion: Organize the trees according to the length of a longest simple path in the tree.)

D27 Show that the average degree in a tree is always less than 2. More specifically, express this average as a function of n, the number of vertices.

D28 Find a tree having the degree sequence $(4, 3, 3, 3, 2, 2, 2, 1, \ldots)$, where the number of vertices of degree 1 is not specified.

D29 Prove that any tree found in problem D28 must have the same number of vertices.

D30 Suppose that a forest with three components has 10 vertices of degree greater than 1, and that these degrees add up to 29. All of the remaining vertices have degree 1. How many vertices of degree 1 must there be?

D31 Show that the number of vertices of degree 1 in any tree must be greater than or equal to the maximum degree in the tree.

D32 (a) Explain how you know that there cannot exist a connected graph having 101 vertices and 99 edges.;

(b) Every connected graph with v vertices must contain at least ____ edges. (Fill in the blank)

D33 Show that if a graph has v vertices and $v - 1$ edges and contains no cycles, then the graph must be connected. (Hint: Let k be the number of components and use a result that was established in this chapter.)

D34 Show that if a graph has v vertices and $v - 1$ edges and is connected, then the graph contains no cycles. (Hint: Since it is connected, the graph contains a spanning tree.)

Three conditions

Problems D33 and D34, along with Tree Theorem 1, show that if a graph **G** satisfies any two of the three conditions;

1. **G** has v vertices and $v - 1$ edges;
2. **G** is connected;;
3. **G** contains no cycles,

then **G** satisfies all three conditions.

D35 Let **P** be a simple path joining vertices A and B in a tree **T**. We will prove that **P** is the only simple path in **T** from A to B by showing that any simple path **Q** in **T** from A to B must contain every edge of **P**.
Suppose UV is an edge of **P** that is not in **Q**. Derive a contradiction by the following steps:

(a) Show that the subgraph $\mathbf{T} - UV$, obtained by removing edge UV from **T**, still contains a path from U to V;

(b) Show that $\mathbf{T} - UV$ contains a simple path from U to V;

(c) Show that $\mathbf{T}$ contains a cycle.

D36 Let $\mathbf{G}$ be a directed graph, not necessarily a tree, that satisfies the indegree condition of the Root Theorem: All but one vertex of $\mathbf{G}$ has indegree 1 and the remaining vertex has indegree 0. Show that any cycle contained in $\mathbf{G}$ must be a directed cycle.

D37 Let $\mathbf{G}$ be a directed graph that satisfies the indegree condition of the Root Theorem as in the preceding problem, and suppose that $\mathbf{G}$ contains no directed cycles. Prove that $\mathbf{G}$ must be a tree. (Suggestion: Use problem D33.)

D38 Suppose that an edge is removed from a tree but its endpoints remain, and then another edge is added, joining two previously nonadjacent vertices.

(a) Find an example of this situation in which the resulting graph is a tree.

(b) Find an example of this situation in which the resulting graph is not a tree.

Cycles and spanning trees

D39 In this graph, the heavy edges form a spanning tree $\mathbf{T}$. For each cycle $\mathbf{C}$ given below, find at least one edge UV in $\mathbf{C}$ that can replace AB in $\mathbf{T}$ so that the resulting subgraph $\mathbf{T} - AB + UV$ is still a spanning tree.

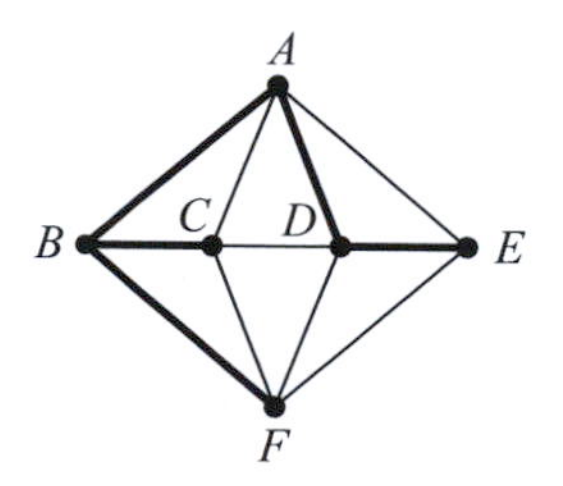

(a) $\mathbf{C}$ is $ABFEA$

(b) $\mathbf{C}$ is $ABCFDA$

(c) $\mathbf{C}$ is $ABCDFEA$

D40 Let $\mathbf{T}$ be a spanning tree in a graph, let $\mathbf{C}$ be a cycle, and suppose that AB is an edge of $\mathbf{C}$ that is also in $\mathbf{T}$. Show that there is at least one other edge UV in $\mathbf{C}$ that can replace AB in $\mathbf{T}$ so that the resulting subgraph $\mathbf{T} - AB + UV$ is still a spanning tree. (Hint: $\mathbf{T} - AB$ is a forest with two components.)

D41 Let $\mathbf{T}$ be a spanning tree in a graph, let $\mathbf{C}$ be a cycle, and suppose that UV is the only edge of $\mathbf{C}$ that is not in $\mathbf{T}$. Show that any other edge AB of $\mathbf{C}$ can be replaced in $\mathbf{T}$ by UV and the resulting subgraph $\mathbf{T} - AB + UV$ is still a spanning tree. (Suggestion: Derive this from the preceding problem.)

D42 A graph $\mathbf{G}$ is formed by removing one edge from $\mathbf{K}_6$. All vertices remain in the graph. Find the number of spanning trees in $\mathbf{G}$.

D43 Find the long and short codewords that correspond to this tree in $\mathbf{K}_5$.

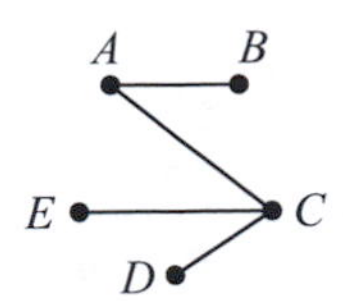

D44 For spanning trees in $\mathbf{K}_n$, the long codeword consists of $2(n-1)$ symbols obtained by appropriate pruning of the tree, and the short codeword consists of the $n-2$ symbols that appear in the even-numbered positions of the long codeword, excluding the last symbol. Find the long and short codewords that correspond to this tree in $\mathbf{K}_6$.

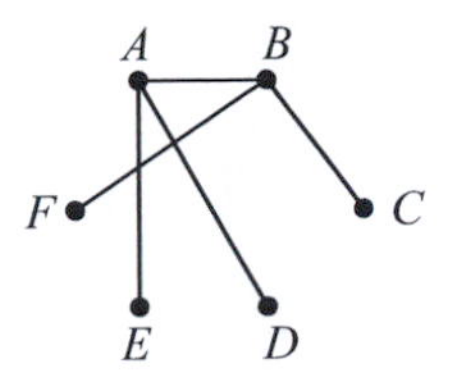

D45 This problem illustrates how to go from the short codeword to the long codeword for a spanning tree in $\mathbf{K}_5$. In this sequence there are three occupied positions (determined by the short codeword), and five empty positions.

$$\_\_ \; \underline{B} \; \_\_ \; \underline{A} \; \_\_ \; \underline{A} \; \_\_ \; \_\_$$

Fill in the five empty positions with the letters A, B, C, D, E, starting at the left, according to the following rules:

1. A letter is available to be placed in an empty position if that letter has not yet been placed in an empty position and if it does not appear in any occupied position that comes later in the sequence;
2. Among the available letters, the first one (in alphabetical order) is used.

Compare your result with that of problem D20.

D46 Find the long codeword for the spanning tree in $\mathbf{K}_5$ that corresponds to each given short codeword:

(a) EBC (b) AAE (c) ECC (d) BAB (e) CCC

D47 Adapt the procedure illustrated in problem D45 to find the long codeword for the spanning tree in $\mathbf{K}_6$ that corresponds to each given short codeword:

(a) $ABCD$ (b) $BAAB$ (c) $CFAC$ (d) $EEEA$

D48 Find the long codeword for the spanning tree in $\mathbf{K}_{10}$ that corresponds to each given short codeword:

(a) $HGCDEFBA$ (b) $CJHHEAIA$ (c) $EEJHDCJA$

Spanning Tree Algorithms

Constructing Spanning Trees

It is often useful to construct a spanning tree in a given connected graph. There are two basic approaches to this problem:

1. **Building up** Start with any edge of the graph and select edges, one at a time, in such a way that at each stage a tree is formed by the selected edges and their endpoints. Continue doing this until the selected edges, together with their endpoints, form a spanning tree.

2. **Reducing down** Start with the entire graph, and if the graph contains any cycles, remove any edge from any cycle. Let the endpoints of that edge remain in the graph. Continue doing this until no cycles remain.

In a connected graph, either of these approaches eventually produces a spanning tree.

E1 How many steps are required by each procedure described above? Answer in terms of v, the number of vertices, and e, the number of edges in the graph.

Weighted Graphs

Below is a graph in which each edge has a number associated with it, which we refer to as the *weight* of that edge. We will refer to such a graph as a *weighted graph.*

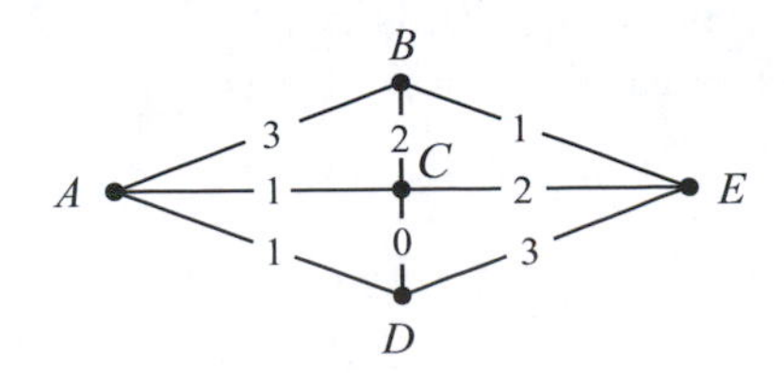

In some applications these numbers might represent cost or travel time.

Minimal Spanning Trees

The *weight of a tree* is defined to be the sum of the weights of all edges in the tree. For example, the weight of this tree is 7.

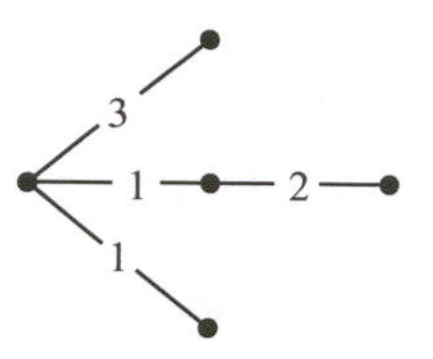

The problem is to find a *minimal spanning tree*, which is a spanning tree having minimum weight among all spanning trees in the graph. The spanning tree shown above is not minimal in the given weighted graph.

E2 Find all of the minimal spanning trees in this weighted graph by trial and error.

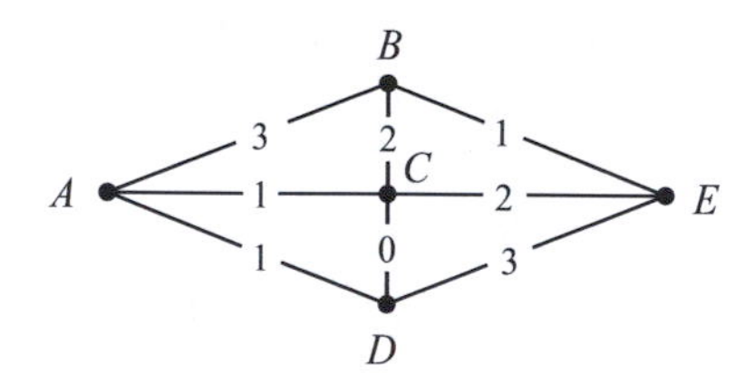

We will give two different procedures for constructing a minimal spanning tree in a given weighted graph. The first procedure constructs the tree by *building up*, the second by *reducing down*.

Prim's Algorithm

Start at any vertex and select an edge having minimum weight among all edges at that vertex. On each step, consider all edges that go from a vertex already reached, to a new vertex. Among these edges, select one that has minimum weight. Continue until all vertices have been reached.

Applying this algorithm in the example above and beginning at A, we find that the edges of minimum weight at that vertex are AC and AD. The algorithm allows either edge to be selected. Arbitrarily selecting AC, we consider vertices A and C to have been reached and look at edges that join one of these vertices to B, D or E.

Among these edges only CD has minimum weight, so it must be selected on this step. We now have $\{A, C, D\}$ as the set of reached vertices and we consider edges that go from one of these to either B or E. This time there are two allowable choices, CB and CE, and we arbitrarily select CB. Finally edge BE is selected.

The sequence of trees produced by these steps is shown below.

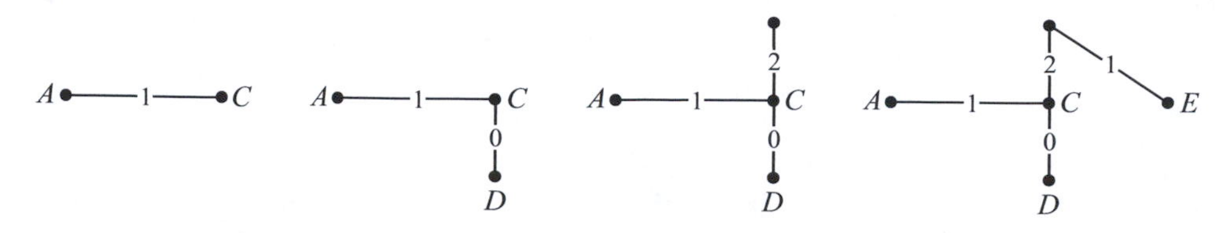

The way in which edges are selected by Prim's Algorithm guarantees that at each stage the selected edges, along with their endpoints, form a tree. When all vertices have been reached, a minimal spanning tree is formed.

E3 Apply Prim's Algorithm to this graph.

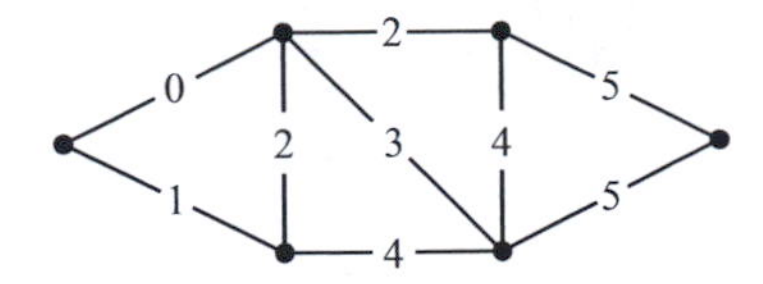

Tables for Prim's Algorithm

Keeping track of all the edges that must be considered at each stage of Prim's Algorithm and comparing their weights can become difficult, especially in a graph with a large number of vertices and edges. Therefore it is usually more convenient to record the relevant information on tables, as illustrated below for the example worked out earlier.

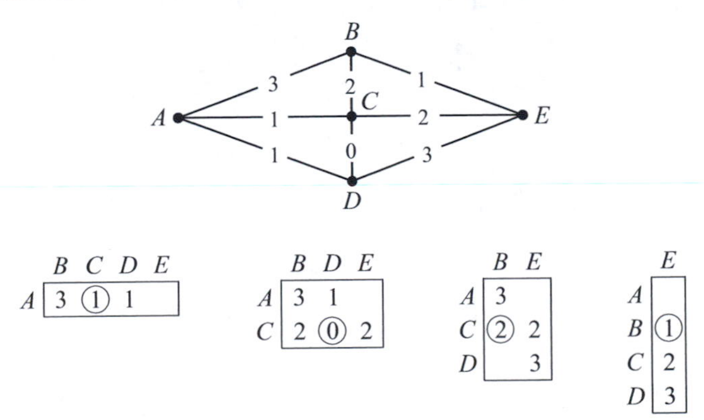

	B	C	D	E
A	3	①	1	

	B	D	E
A	3	1	
C	2	⓪	2

	B	E
A	3	
C	②	2
D		3

	E
A	
B	①
C	2
D	3

The weights in each table correspond to the edges that were considered on each step. The circled weights indicate the particular choices that were made.

Next we look at a different approach to the construction of a minimal spanning tree.

The Reduction Algorithm

Start with the entire graph. On each step select any cycle and remove any edge that has maximum weight among the edges of that cycle. Continue until no cycles remain.

Returning again to this example

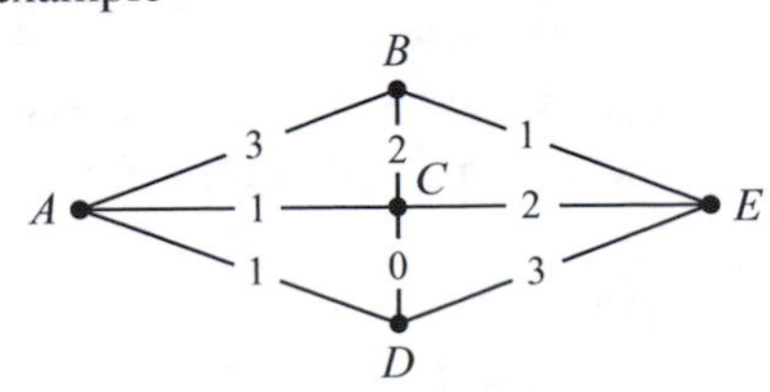

and applying the Reduction Algorithm, we find a number of choices for the first edge to be removed.

E4 There are only two edges in this graph that cannot be removed on the first step. Which are they?

Selecting edge AB and removing it, we reduce to the graph

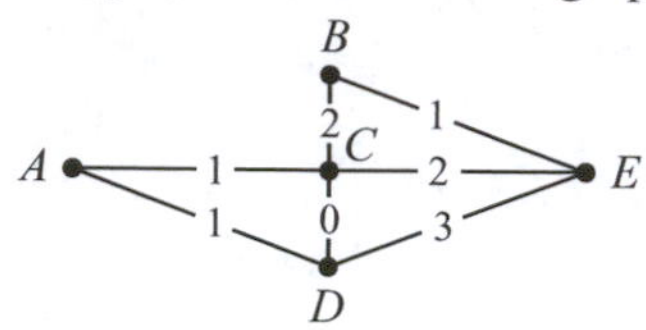

E5 Carry out the remaining steps of the Reduction Algorithm for this graph.

E6 Apply the Reduction Algorithm to the graph in problem E3.

Spanning Trees and Shortest Paths

Spanning trees play an essential role in the solution of the *shortest path problem*. Consider the problem of finding a shortest path from A to B in this graph.

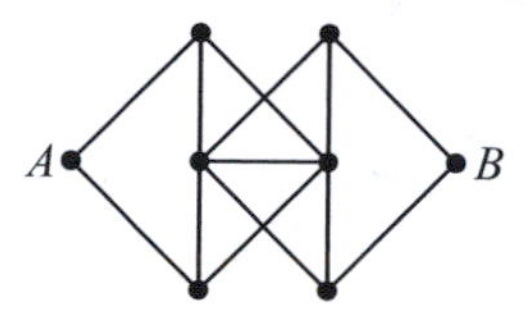

The first edge along such a path must go from A to an adjacent vertex, so we begin by including all such edges.

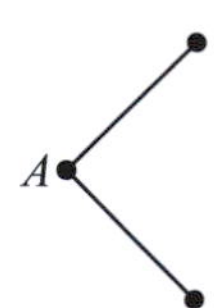

Also, since a path follows its edges in a particular direction, it seems reasonable to place directions on these edges.

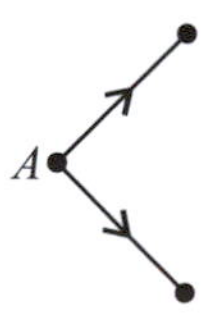

In other words, we are growing a *directed tree* starting at vertex A. Recall from chapter D that a *root* for a directed tree is a vertex from which all other vertices of the tree are reachable; so vertex A is a root for this tree.

In the next step we include edges that go to new vertices from vertices already reached, with appropriate directions. However only one edge is added going to each new vertex, so the construction remains a tree.

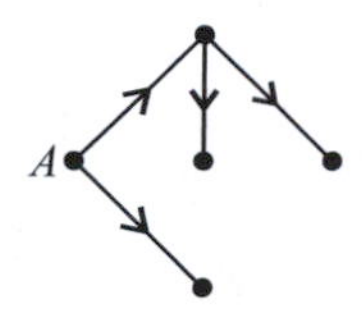

Continuing in this way for two more steps, we end up with a spanning directed tree **T** in which A is a root.

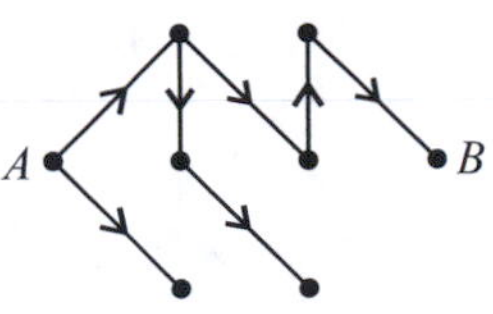

E7 Verify that for every vertex X in this graph, **T** contains a shortest path from A to X.

Thus the construction of **T** not only solves the shortest path problem from A to B, but it simultaneously solves all shortest path problems starting at A.

Minimal Paths in a Weighted Graph

A generalization of the shortest path problem is the problem of finding a simple path of minimum weight (a *minimal path*) joining two given vertices, where we define the *weight of a path* to be the sum of the weights of all edges in the path. This is the *minimal path problem*.

E8 By trial and error, find minimal paths in this weighted graph from vertex A to each of the other vertices.

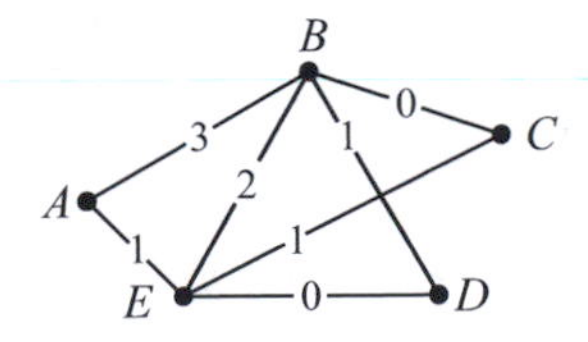

As with shortest paths, there is always a spanning tree that contains minimal paths from a given vertex to every other vertex in a weighted graph. To construct such a tree, a reasonable approach might be to grow one as in Prim's algorithm, as follows:

Minimal Path Algorithm, first attempt

Start with an edge of minimum weight at A, directing it outward from A, and on each step consider the edges that go from a vertex already reached, to a new vertex. Among these edges, select one that has minimum weight. Continue until all vertices have been reached.

E9 Try that for this graph. What goes wrong?

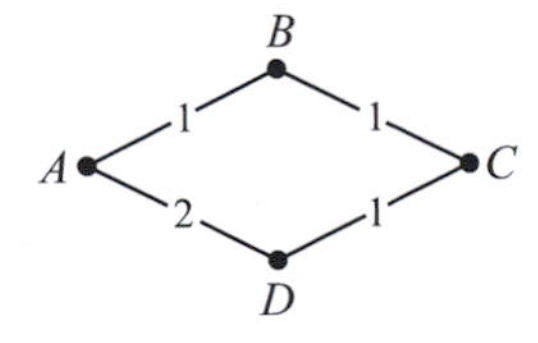

As this example shows, we need to revise our strategy. The problem occurred on the third step, following the construction of this directed tree:

B

A C

While edge CD has minimum weight among the two candidates for inclusion in the tree on the third step, it is not the one that produces the minimal path from A to D. The situation is clarified by the inclusion of *vertex labels* indicating the weight of the path from A to each vertex in the tree.

The next edge selected produces a label at a new vertex, necessarily D in this case. The selection of CD produces the label 3 at D, while the selection of AD produces the label 2. This comparison suggests the new strategy:

Minimal Path Algorithm, revised

Start by placing the label 0 at vertex A and selecting an edge of minimum weight at A, directing it outward from A. Whenever a new vertex X is reached, place a label indicating the weight of the path from A to X in the tree. On each step, consider the edges that go from a vertex already reached, to a new vertex. Among these edges, select one that results in the smallest possible label at the new vertex. Continue until all vertices have been reached.

This procedure, known as *Dijkstra's Algorithm*, produces a spanning directed tree that contains minimal paths from a vertex A to every other vertex in a given weighted graph. We will refer to such a tree as an *optimal A-rooted spanning tree.*

E10 Apply Dijkstra's Algorithm to construct an optimal A-rooted spanning tree in this graph.

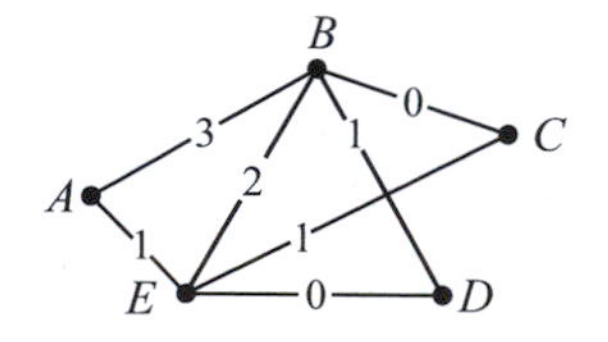

Tables for Dijkstra's Algorithm

Dijkstra's Algorithm can be carried out more easily if we keep track of the necessary information on tables, as we did with Prim's Algorithm. We illustrate this by constructing an optimal A-rooted spanning tree in this graph.

Clearly the first edge is AC. The first tree, with vertex labels, is this:

To determine the next edge, we construct the table

		B	D	E	F
A	0	2			
C	1	1		1	

in which the first column shows the vertex labels at vertices already reached, and the remaining columns show the weights of all edges that are candidates for inclusion on the next step. By adding each vertex label to all weights in the same row, we obtain the values that have to be compared in order to determine which edge to add.

	B	D	E	F
A	2			
C	2		2	

This shows that any one of the edges AB, CB or CE can be added on the next step. Arbitrarily selecting AB, we obtain this labeled tree and the corresponding tables:

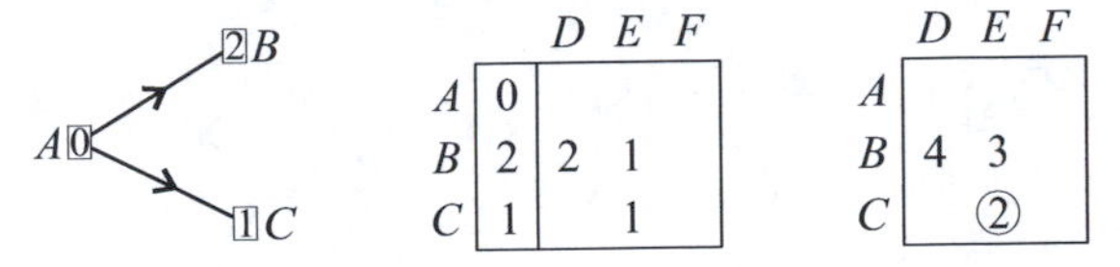

This time the only choice is CE, corresponding to the minimum value on the last table. The remaining steps of the algorithm are shown below, with the circled entry indicating the selected edge and the new vertex label on each step.

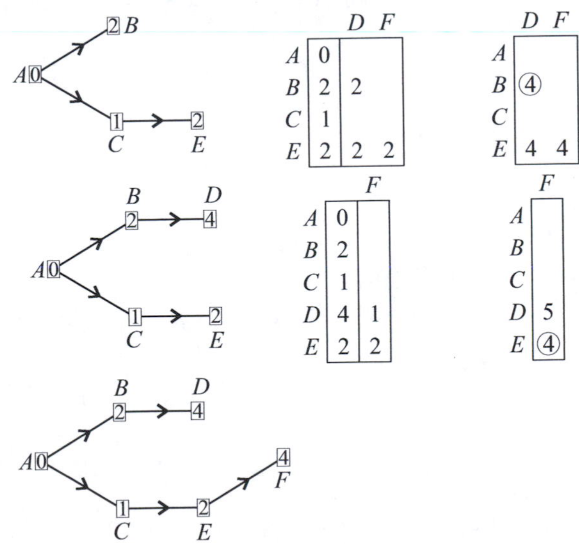

E11 Use the table method to construct an optimal A-rooted spanning tree in this graph.

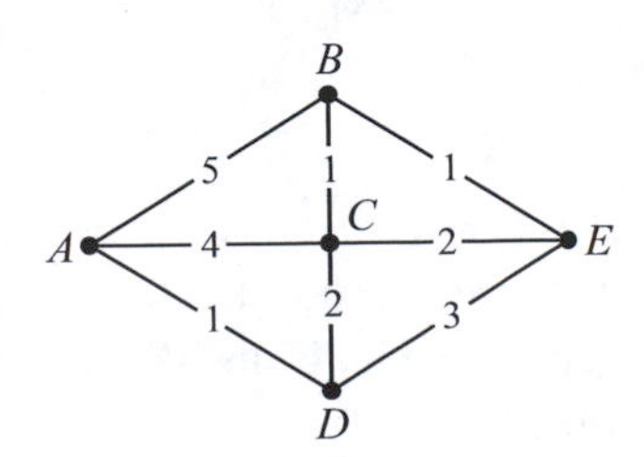

Minimal Paths in a Directed Graph

When some or all of the edges of a weighted graph are directed, paths are required to follow the given directions on all directed edges.

E12 By trial and error, find a shortest path from A to B in this mixed graph.

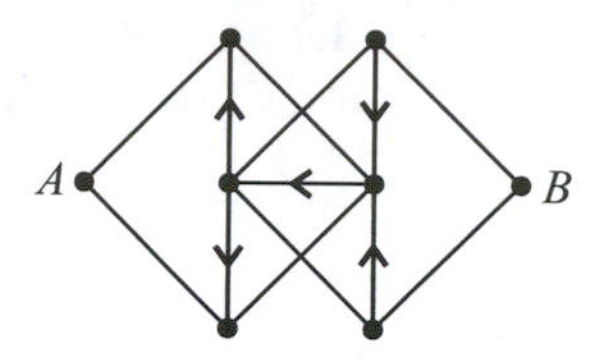

As in the undirected case, we can solve this problem by constructing an A-rooted spanning tree. The sequence of steps is shown below.

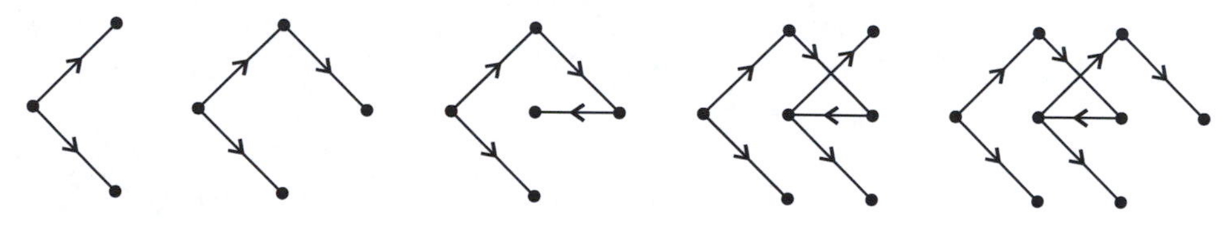

The presence of directed edges affects Dijkstra's Algorithm by restricting the choice of allowable edges on some steps, but otherwise the process is the same. For example, in this weighted mixed graph

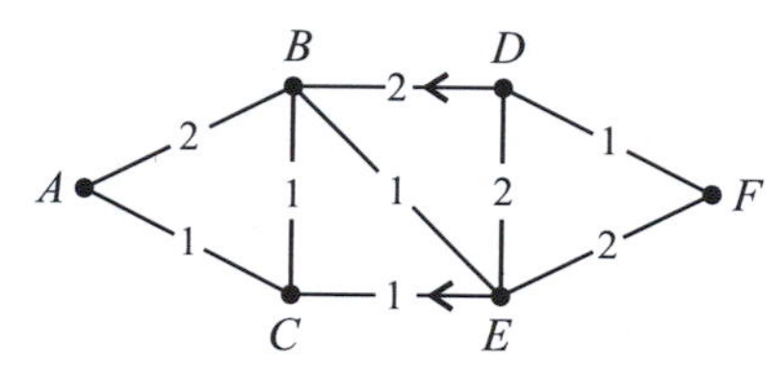

the first two steps of Dijkstra's Algorithm are

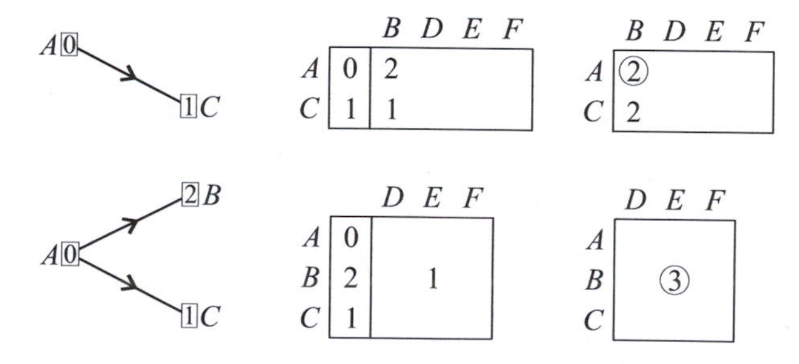

E13 Continue this process, constructing an optimal A-rooted spanning tree.

Negative Weights

Up until now, we have considered only weighted graphs in which all weights are greater than or equal to 0. However there are realistic situations in which we might want to allow some weights to be negative. For example, if the weights represent costs, then a negative weight would represent a profit.

E14 There are two A-rooted spanning trees in this weighted directed graph.

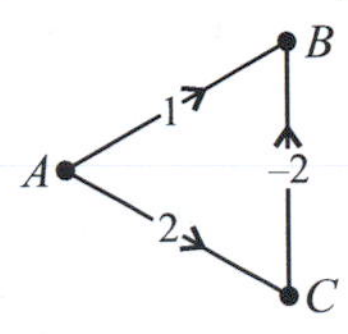

(a) Which one is produced by Dijkstra's Algorithm?

(b) Which one is optimal?

This example illustrates the limitation on Dijkstra's Algorithm: when negative weights occur, the algorithm cannot be counted on to produce an optimal spanning tree. However, there is a simple process that provides a way around this problem. The idea is to introduce a series of *improvement steps* in which the edges of a spanning tree are replaced by other edges that allow vertex labels to be reduced. In the example above, we would improve the A-rooted spanning tree

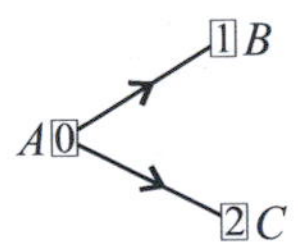

by replacing edge AB with edge CB. The rationale for doing this is that edge CB provides a better route to vertex B, allowing the label at B to be reduced from 1 to 0:

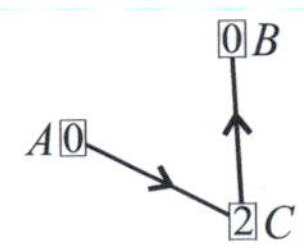

In general, an *improvement step* adds an edge UV to a rooted directed tree whenever the weight w of that edge and the two vertex labels $L(U)$ and $L(V)$ satisfy the condition $\boldsymbol{L(U) + w < L(V)}$. In that case, the label at V is reduced to $L(U) + w$, and UV replaces the existing incoming edge at V. The reduction of the label at V creates a chain reaction causing the labels at all vertices reachable from V in the tree to also be reduced.

E15 Construct an optimal A-rooted spanning directed tree in this graph

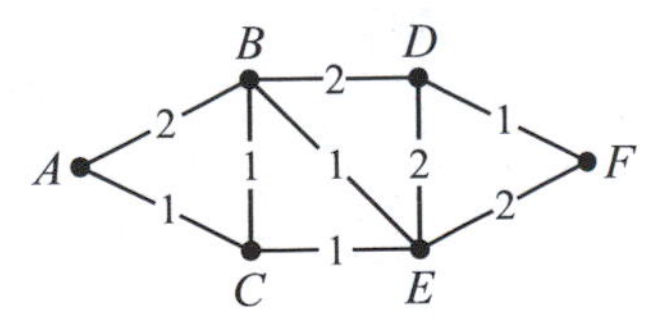

by applying improvement steps to this tree. Begin by inserting vertex labels.

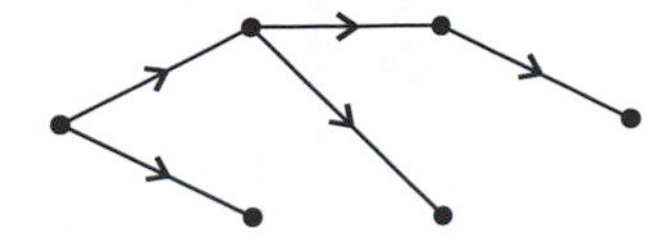

The process illustrated by these last two examples is known as *Ford's Algorithm*. Starting with a rooted spanning directed tree in a weighted graph, improvement steps are applied until the tree becomes optimal.

Ford's Algorithm provides an alternative to Dijkstra's Algorithm that can be used when negative weights occur, as well as when they don't. Dijkstra's Algorithm applies only when all weights are nonnegative.

E16 Use Ford's Algorithm to construct an optimal A-rooted spanning directed tree in this slightly mixed graph

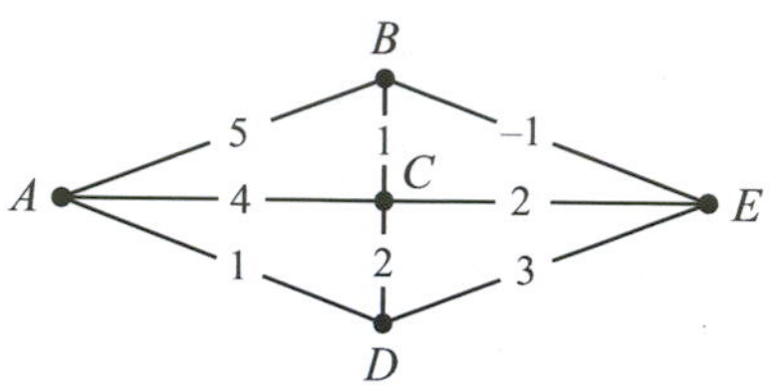

starting with this tree.

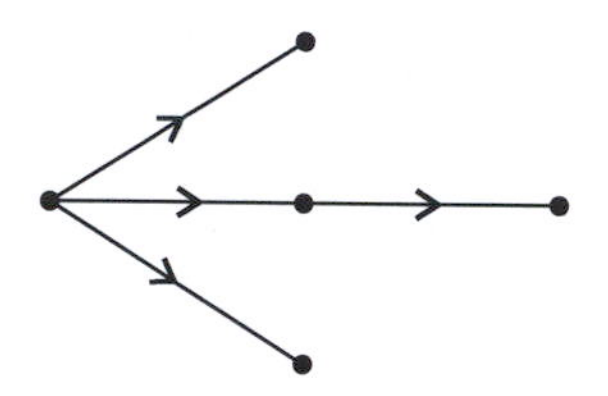

E17 See what happens when Ford's Algorithm is applied to this graph, starting with the given tree.

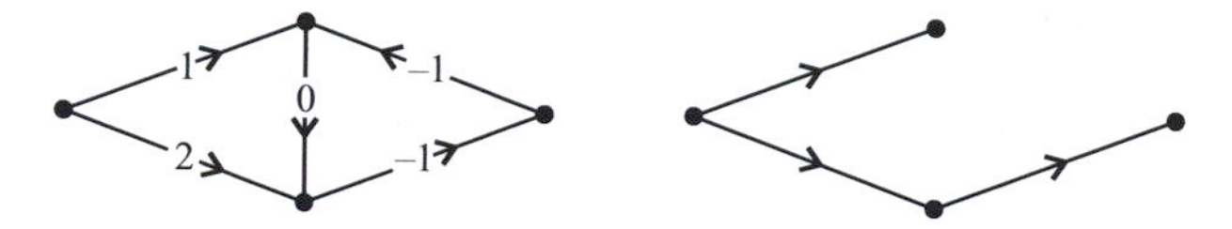

E18 Show that no optimal rooted spanning directed tree exists in the graph of problem E17.

This example illustrates what can go wrong when Ford's Algorithm is applied in a weighted graph that includes negative weights: If the graph contains a cycle having total weight less than zero (a *negative cycle*), then the process is not guaranteed to produce an optimal rooted spanning directed tree. On the other hand, if no negative cycles occur in the graph and all vertices are reachable from a fixed starting vertex A, then the algorithm always results in an optimal A-rooted spanning directed tree in a finite number of steps. The next two problems suggest the main ideas that go into the justification of Ford's Algorithm.

E19 Assume that this configuration is part of a weighted graph that contains no negative cycles.

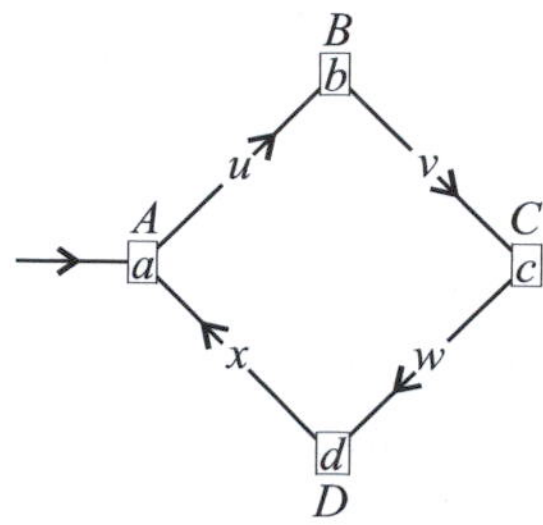

Suppose that a rooted spanning tree includes the edge coming into A from the left, and also edges AB, BC, and CD. Vertex labels a, b, c, d and weights u, v, w, x are as indicated. Therefore, $a + u = b$, $b + v = c$, and $c + w = d$. Show that Ford's Algorithm would not add edge DA to the tree on the next step. In other words, prove that $d + x \geq a$.

Problem E19 shows why Ford's Algorithm continues to maintain a rooted spanning tree on all steps: It never adds an edge that would complete a cycle. This will be explained more completely in problem E40.

The next problem shows that, at least when all weights are integers, the algorithm terminates in a finite number of steps.

E20 Suppose that Ford's Algorithm is applied in a weighted graph in which all weights are integers and no negative cycles occur.

(a) Show that on each step, the sum of all of the vertex labels decreases.

(b) Explain why the algorithm cannot go on for an infinite number of steps.

More Problems

E21 Use the table method to carry out Prim's Algorithm for the graph in problem E3.

E22 In the graph of problem E4, which edges can be selected on the first step of Prim's Algorithm?

E23 In the graph of problem E3, which edges can be removed on the first step of the Reduction Algorithm?

E24 Find a spanning tree of maximum weight in this graph

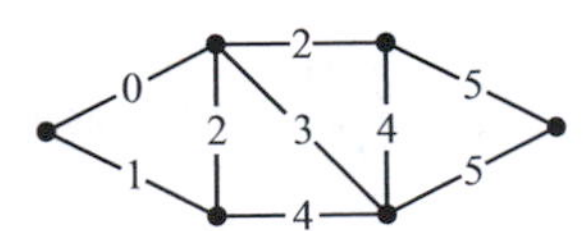

(a) by trial and error;

(b) by applying the Reduction Algorithm with *minimum* in place of *maximum*;

(c) by applying Prim's Algorithm with *maximum* in place of *minimum*.

E25 Construct an R-rooted spanning tree containing shortest paths from R to all other vertices in this graph.

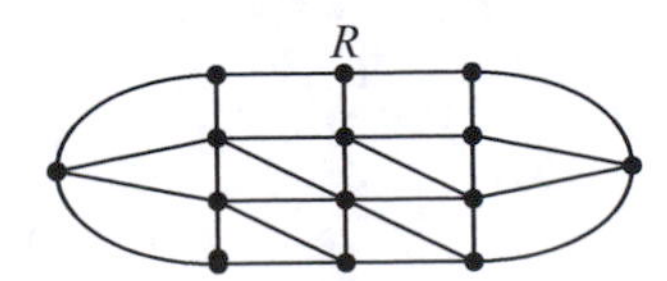

E26 Construct an R-rooted spanning tree containing shortest paths from R to all other vertices in this mixed graph.

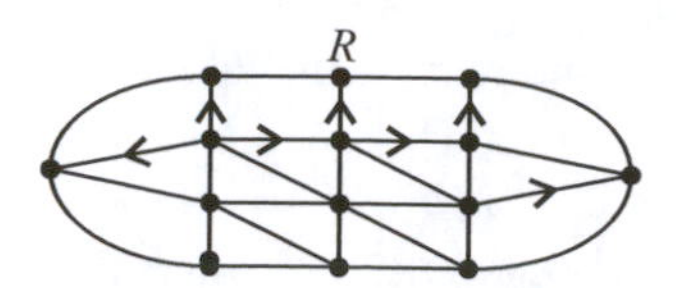

E27 After five steps, Dijkstra's Algorithm produces the R-rooted tree indicated by the heavy edges in this weighted graph. Complete the process, carrying out the final four steps

(a) without using the table method;

(b) using the table method.

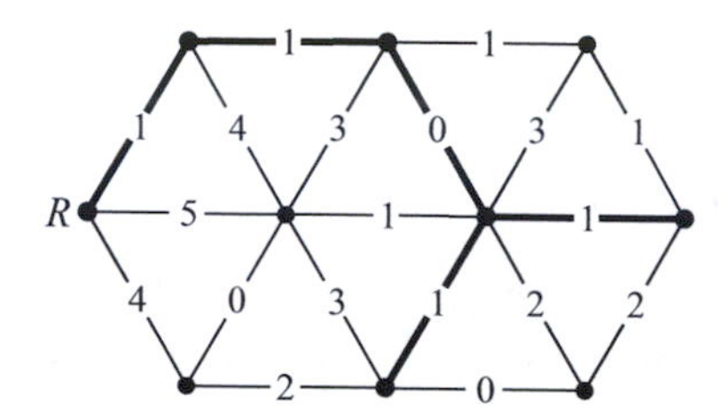

E28 Repeat problem E27 for this mixed graph with directions as shown on three edges.

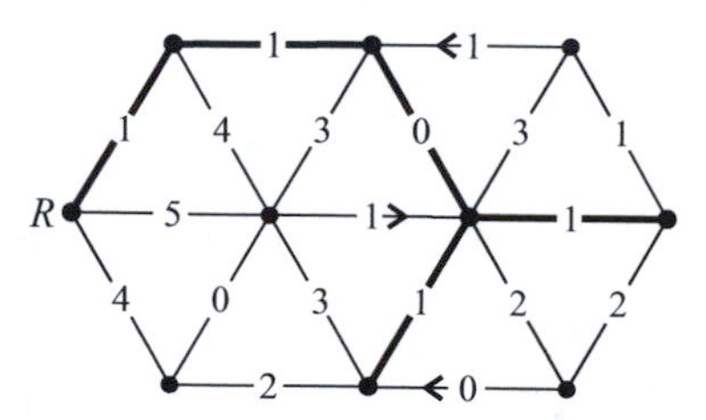

E29 Use Ford's Algorithm to construct an optimal R-rooted spanning tree in this weighted graph starting with the tree at the right.

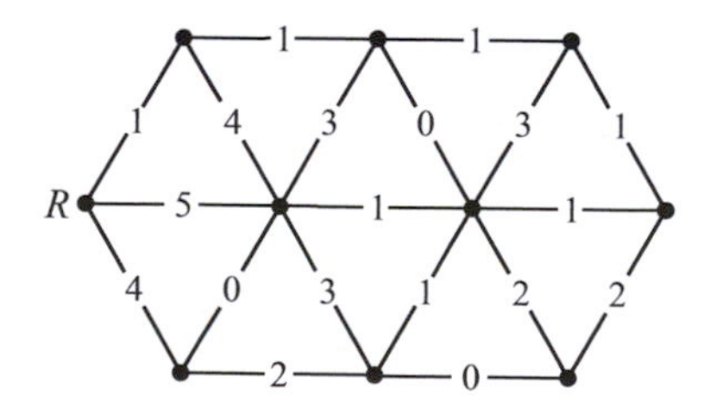

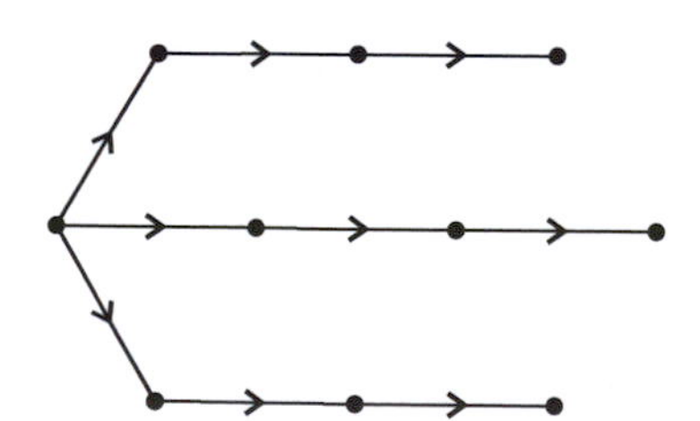

E30 Use Ford's Algorithm to construct an optimal R-rooted spanning tree in this weighted directed graph

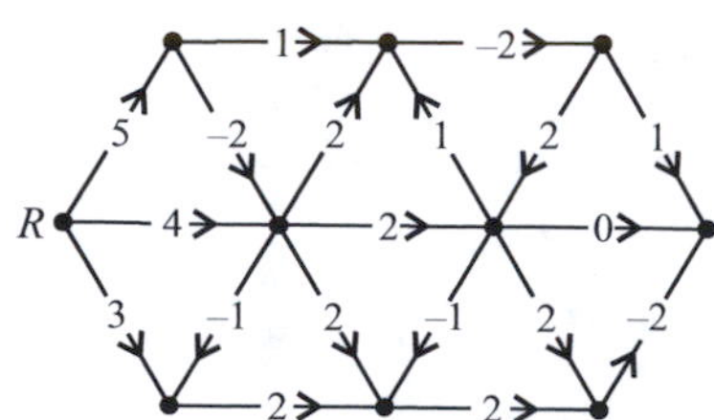

starting with the same tree as in the preceding problem.

Justification of the reduction algorithm

E31 Let AB be an edge of maximum weight in some cycle C in a weighted connected graph G.

(a) Suppose that **T** is a spanning tree in **G** containing AB. Use problem D40 to show that there is another spanning tree that has weight less than or equal to that of **T** and does not contain AB.

(b) Explain why the Reduction Algorithm produces a minimal spanning tree in **G**. (Suggestion: Let **T** be a minimal spanning tree and suppose that **T** contains AB. Then what else exists?)

E32 Suppose that AB is the unique edge of maximum weight in some cycle C in a weighted connected graph **G**. (i.e., the weight of AB is strictly greater than that of any other edge in C.) Show that AB is not contained in any minimal spanning tree in **G**.

Justification of Prim's Algorithm

E33 Let UV be an edge of minimum weight at some vertex U in a weighted connected graph **G**.

(a) Suppose that **T** is a spanning tree in **G** not containing UV. Show that there is another spanning tree in **G** that contains UV and has weight less than or equal to that of **T**. (Hint: **T** contains a simple path from V to U. Combine this with UV to form a cycle and apply problem D41 with $A = U$.)

(b) Show that UV is contained in some minimal spanning tree in **G**.

E34 Let $\mathbf{T}_0$ be the tree formed by the first m steps of Prim's Algorithm in a weighted connected graph **G**, and let UV be the next edge selected by the algorithm. Suppose that **T** is a minimal spanning tree in **G** that contains $\mathbf{T}_0$ but not UV. Show that there is another minimal spanning tree in **G** that contains $\mathbf{T}_0$ and also UV.

E35 Explain how the two preceding problems show that Prim's Algorithm produces a minimal spanning tree in **G**.

E36 Suppose that UV is the unique edge of minimum weight at some vertex U in a weighted connected graph **G**. Show that UV must be contained in every minimal spanning tree in **G**.

E37 Find an example of a weighted connected graph **G** that contains an edge AB having the properties

1. AB is the unique edge of minimum weight in some cycle, and
2. AB is not contained in any minimal spanning tree in **G**. (Hint: Arrange for AB to be in more than one cycle.)

Justification of Dijkstra's Algorithm

Starting at a vertex R in a weighted graph in which all weights are nonnegative, Dijkstra's Algorithm produces an R-rooted spanning tree **T**, which we claim is optimal. This means that every directed path from R to any other vertex X has total weight greater than or equal to the label at X, which is the weight of the simple path from R to X in $\mathbf{T}$.

E38 Consider this situation, in which a, b, c are vertex labels at the end of Dijkstra's Algorithm and u, v, w are weights.

$$\underset{R}{\boxed{0}} \xrightarrow{u} \underset{A}{\boxed{a}} \xrightarrow{v} \underset{B}{\boxed{b}} \xrightarrow{w} \underset{C}{\boxed{c}}$$

Suppose that $u \geq a$, $u + v \geq b$, and $u + v + w < c$. We will show that this is impossible by deriving a contradiction.
Consider the order in which the vertices A, B and C are reached in the construction of $\mathbf{T}$.

(a) Show that A must have been reached before C. (Suggestion: If not, then edge RA would have been a candidate for inclusion in $\mathbf{T}$, but not selected, on the step in which C was reached. This tells you something about the relationship between c and u. What is it, and how is that a contradiction?)

(b) Use the result of (a) to show that B must also have been reached before C. (Suggestion: If not, imitate the argument in (a) using edge AB in place of RA and show that $c \leq a + v$. Again derive a contradiction.)

(c) Use (b) to show that $c \leq b + w$ and derive a contradiction.

(d) Where did you use the condition that all weights are nonnegative?

E39 Generalize problem E38 to show that any directed path starting at R has total weight greater than or equal to the label at its final vertex. (Suggestion: If not, then among all paths that fail to have this property, consider one having the fewest edges.)

Justification of Ford's Algorithm

Assume that Ford's Algorithm is applied starting with an R-rooted spanning tree in a weighted graph in which no negative cycles occur. We claim three things:

1. The algorithm maintains an R-rooted spanning tree on all steps.

2. The algorithm terminates in a finite number of steps.

3. After the final step, the tree is optimal.

E40 Proof of (1):

(a) Generalize problem E19 to show that the algorithm never adds an edge that would complete a directed cycle.

(b) Use problem D37 to show that the algorithm maintains an R-rooted spanning tree on all steps.

E41 Proof of (2): Generalize problem E20 to show that the conclusion still holds (the algorithm cannot go on for an infinite number of steps) when the weights are assumed only to be real numbers, not necessarily integers. (Hint: The result in E20(a) is still valid, and there are only finitely many possible spanning trees in the graph.)

E42 Proof of (3): Again, the optimality condition means that every directed path starting at R has total weight greater than or equal to the label at its final vertex.

(a) Consider this situation, in which a, b, c are vertex labels at the end of Ford's Algorithm and u, v, w are weights. Prove that $u + v + w \geq c$.

$$\underset{\boxed{0}}{R} \xrightarrow{u} \underset{\boxed{a}}{A} \xrightarrow{v} \underset{\boxed{b}}{B} \xrightarrow{w} \underset{\boxed{c}}{C}$$

(Suggestion: First show that $u \geq a$, $a + v \geq b$, and $b + w \geq c$.)

(b) Generalize (a) to show that every directed path starting at R has total weight greater than or equal to the label at its final vertex.

Euler Paths

The Königsberg Bridge Problem

In the city of Königsberg there were seven bridges connecting two islands and portions of the city on each bank of a river. The residents would try to walk in such a way as to cross each bridge exactly one time and return to the starting point.

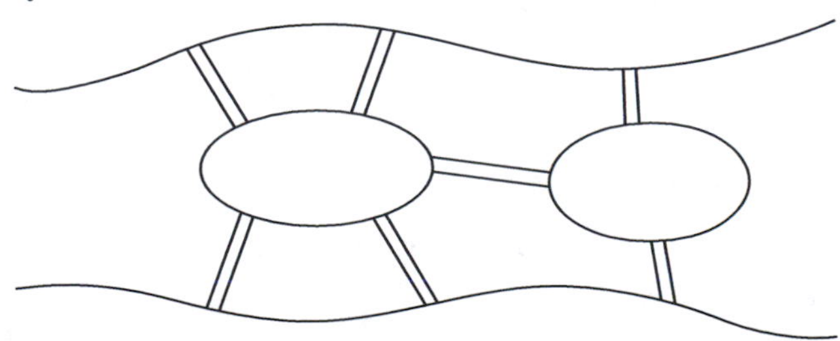

No one was ever able to do this, and Euler explained why in a 1736 paper that is considered to be the beginning of graph theory.

The same problem can be restated in graph-theoretic terms by reducing each of the four regions down to a vertex: In this multigraph we are looking for a closed path that includes each edge exactly once.

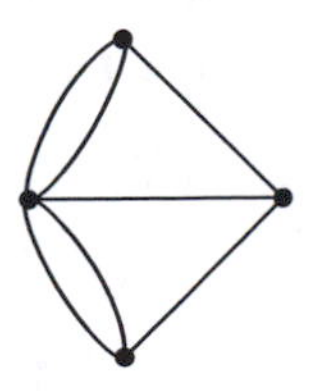

Definitions An *Euler path* in a graph or multigraph **G** is a path that includes every edge of **G** exactly once. (Recall that in a path, vertices can be repeated. In general so can edges, but not in an Euler path.) An Euler path is called *closed* if it starts and ends at the same vertex; otherwise it's called *open*.

F1 Find an Euler path, either closed or open, in each graph or multigraph below, and notice where odd vertices occur in each case.

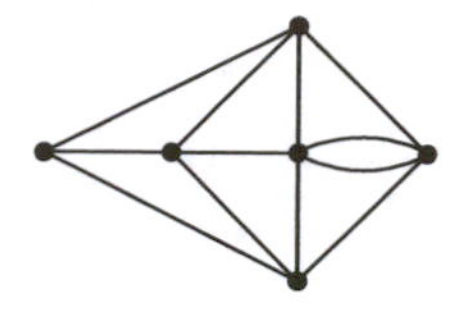 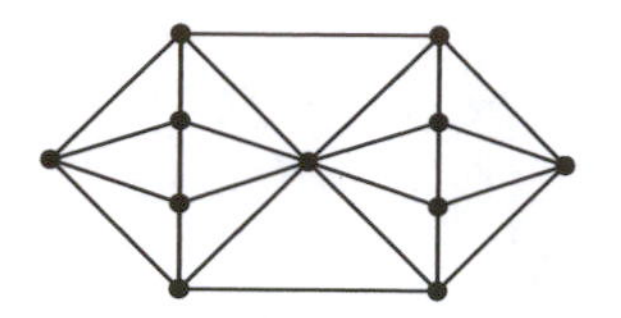 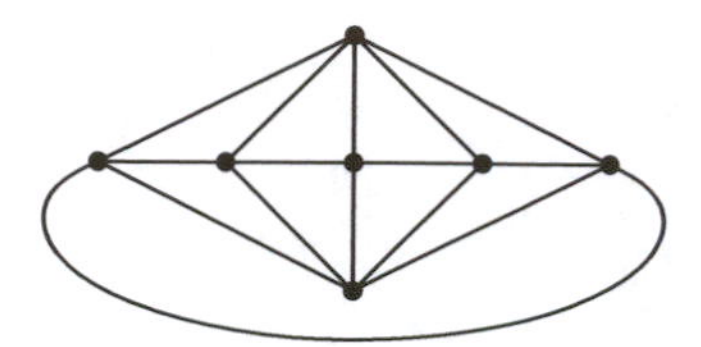

F2 Explain why a graph or multigraph that has more than two odd vertices cannot contain an Euler path.

Returning to the Königsberg Bridge Problem, we can now see why there is no way to cross each bridge exactly once, whether or not the walk ends at its starting point: There are too many odd vertices.

F3 What can you say about the number of odd vertices in a graph or multigraph that contains

(a) a closed Euler path?
(b) an open Euler path?

F4 (a) Can a disconnected graph or multigraph contain an Euler path?
(b) Take into account the possibility that there may be vertices of degree 0. Does that change your answer to (a)?

In general, the question of whether a given graph contains an Euler path, either closed or open, is answered by the following theorem:

The Euler Path Theorem

1. *A graph or multigraph contains a closed Euler path if and only if*
   (a) *Every vertex has even degree, and*
   (b) *All edges are in the same component.*
2. *A graph or multigraph contains an open Euler path if and only if*
   (a) *Exactly two vertices have odd degree, and*
   (b) *All edges are in the same component.*

F5 How many more bridges would have to be built in Königsberg so that it becomes possible to cross each bridge exactly once and return to the starting point?

F6 For each multigraph below, either find an Euler path or else indicate how you know it doesn't exist.

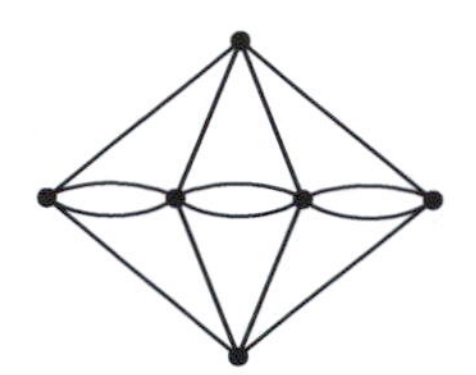  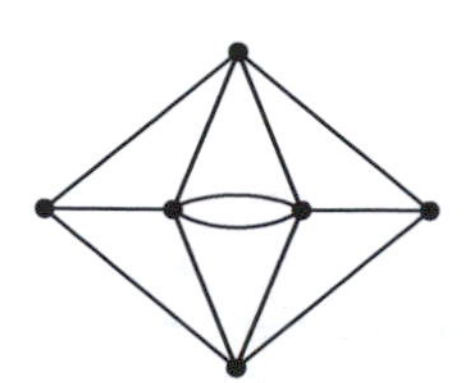

F7 For which n-values does the complete graph $\mathbf{K}_n$ contain a closed Euler path? Does any $\mathbf{K}_n$ contain an open Euler path?

The Euler Path Theorem says that conditions (a) and (b) are both necessary and sufficient for the existence of a particular type of path. The fact that they are sufficient — that is, that they guarantee the existence of an Euler path — is the more difficult part to prove, and the following problems give some indication of how this is done.

F8 Show that if a graph has at least one edge, and if all vertices have even degree, then the graph must contain at least one cycle. (Hint: Suppose it doesn't. Then any component that contains an edge is a tree with at least two vertices. If necessary, refer back to chapter D to find a contradiction.)

F9 Construct a closed Euler path in this graph by the following procedure:

1. Remove the edges of the cycle $ABCDEFGHA$.
2. In each component of the remaining graph, find a closed Euler path.
3. Combine these paths with the cycle.

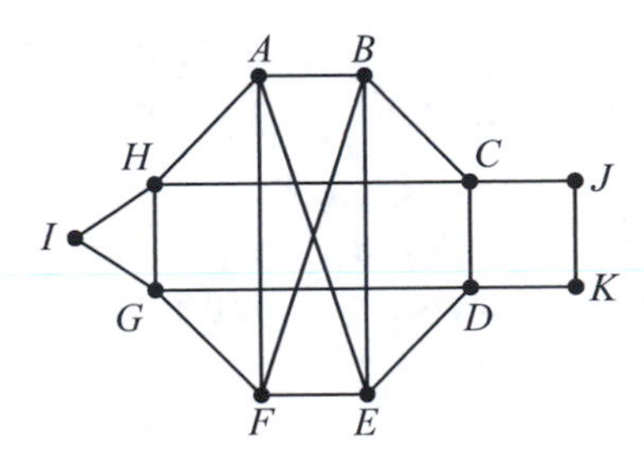

The significance of problems F8 and F9 is that in any graph satisfying conditions (a) and (b) in part (1) of the theorem, a closed Euler path can be built up by a series of steps in which a cycle is combined with closed Euler paths from subgraphs. Another way to say this is that the construction in problem F9 represents the induction step in a proof by mathematical induction on the number of edges in the graph.

Problem F20 will show how part (1) of the Euler Path Theorem can be used to prove part (2).

Euler Paths in Directed Graphs and Directed Multigraphs

When each edge in a graph or multigraph has a particular direction associated with it, an Euler path is required to follow these directions.

F10 Find an Euler path in this directed multigraph.

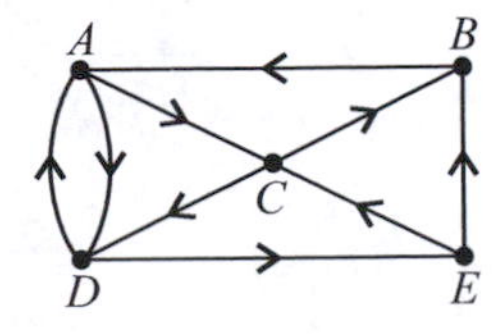

F11 In a directed graph or multigraph, we consider the indegree and outdegree at each vertex. For example, at vertex B above, the indegree is $d_{\text{in}}(B) = 2$, while the outdegree is $d_{\text{out}}(B) = 1$. What do you notice about the indegrees and outdegrees in this example that makes it possible for an Euler path to exist?

F12 Fill in the missing conditions below.

Euler Path Theorem, directed version

1. *A directed graph or multigraph contains a closed Euler path if and only if*
   (a) *At each vertex,* ________, *and*
   (b) *All edges are in the same component.*
2. *A directed graph or multigraph contains an open Euler path if and only if*
   (a) *At one vertex,* ________,
   at another vertex, ________,
   at every other vertex, ________, *and*
   (b) *All edges are in the same component.*

Application of Euler Paths: State diagrams, DeBruijn sequences, and rotating wheels

A *state diagram* consists of a directed graph or multigraph in which each vertex is labeled with a binary string (a sequence of 0's and 1's) representing the different possible states of a system, and the edges represent transitions between states. We are interested here in state diagrams in which each transition removes the first bit (0 or 1) from the string and creates a new state by adding a bit at the end. The added bit is represented as a label on the edge. For example,

$$\boxed{10010} \xrightarrow{0} \boxed{00100}$$

F13 Fill in the missing states and labels on edges:

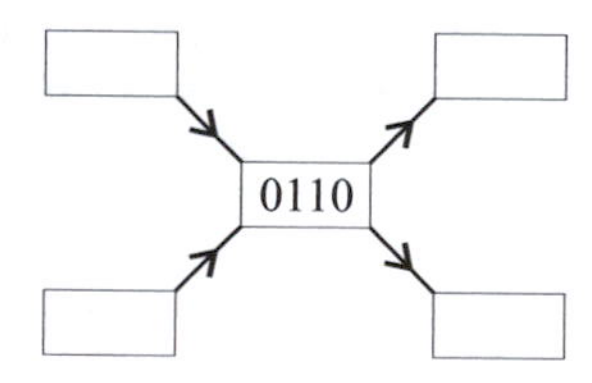

F14 Insert edges, with labels, in the appropriate places in this state diagram. Two of the edges are loops (edges that start and end at the same vertex).

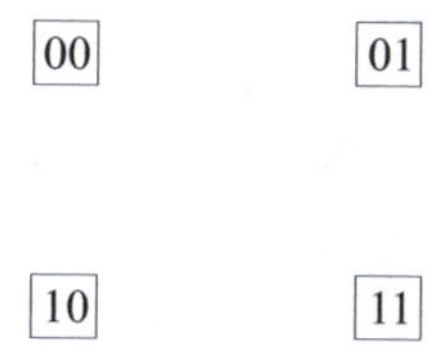

F15 Find an Euler path in the state diagram above. What property of the diagram guarantees that such a path exists?

By following the Euler path obtained in problem F15, we can form a binary string of length 8 consisting of the edge labels along the path. The result is either 00011101 or 11100010, or a rotation of one of these. (By that we mean a string obtained by moving a segment from the beginning of the string to the end. For example, 01110100 is a rotation of 00011101.)

F16 Using the string of length 8 that you obtained from the Euler path found in problem F15, list all of the binary triples (strings of length 3) that are segments of the string, starting at each of the eight positions. At the end of the string, go back to the beginning when necessary to form the triples. What do you notice about these eight triples?

The string of length 8 used in the problem above is an example of a *DeBruijn sequence* based on the set $\{0, 1\}$. In general, a Debruijn sequence of length 2^n based on $\{0, 1\}$ has the property that the segments of length n, starting at each of the 2^n positions, are all different, and then (automatically) must include all possible strings of n 0's and 1's.

F17 Explain why the string of length 8 obtained from any Euler path in problem F15 is guaranteed to be a DeBruijn sequence. Do this by showing that every binary triple abc occurs as a segment somewhere in the string. (Hint: Go to

$$\boxed{ab}\xrightarrow{c}$$

in the state diagram and follow the Euler path back two steps.)

As a practical application of a DeBruijn sequence, consider a rotating wheel, around which a sequence of eight switches are arranged. Each switch is either on or off, as represented in the diagram by a 1 or a 0. Different positions of the drum cause different 3-bit sequences to appear in a particular location where they connect with incoming wires. Because the bits are arranged in a DeBruijn sequence, every possible on-off triple can be attained by some position of the wheel.

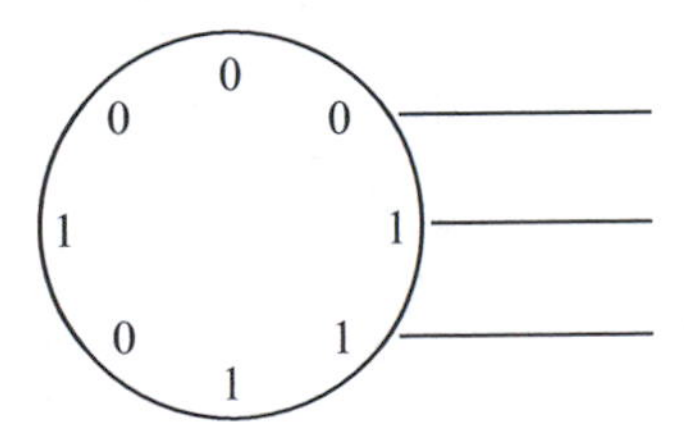

More generally, 2^n switches are arranged around a wheel in a DeBruijn sequence to connect with n incoming wires.

More Problems

F18 (a) Add the smallest possible number of edges to this graph in order to produce a graph that contains an open Euler path.

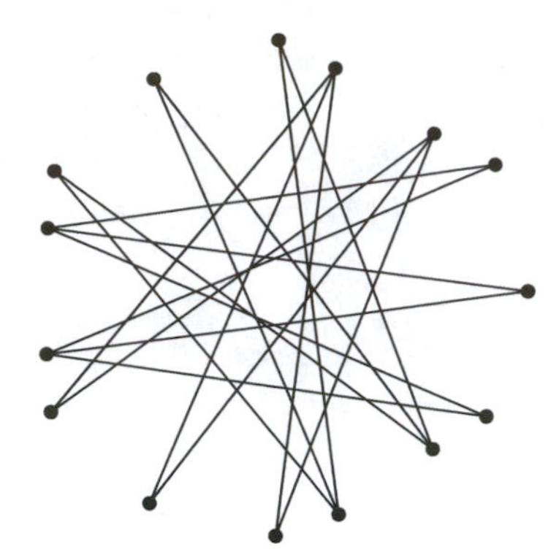

(b) Do the same for a closed Euler path.

F19 Is it possible to draw this figure in one continuous motion? This means without lifting your pen from the paper.

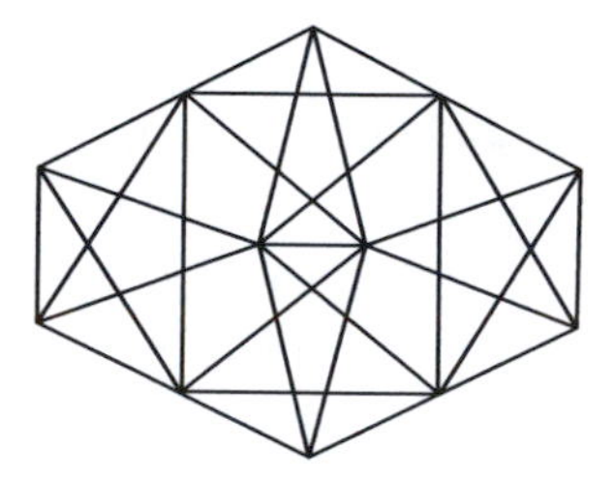

F20 Suppose that a graph or multigraph **G** satisfies conditions (a) and (b) in part (2) of the Euler Path Theorem, and assume that part (1) has been proved. Prove that **G** contains an open Euler path. (Hint: Add one edge.)

F21 Find an Euler path in this partially directed multigraph.

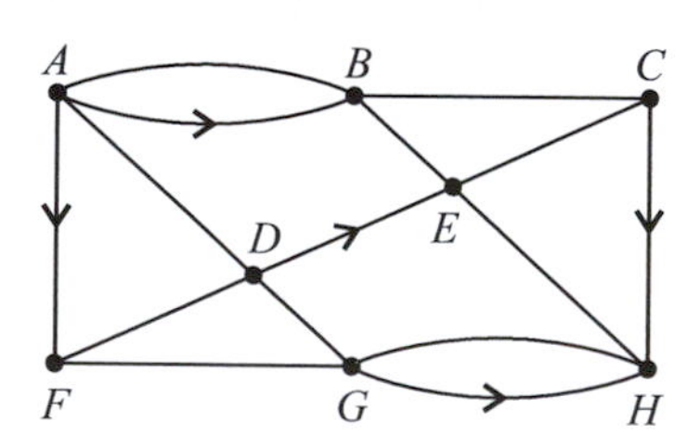

F22 Show that no Euler path exists in this one.

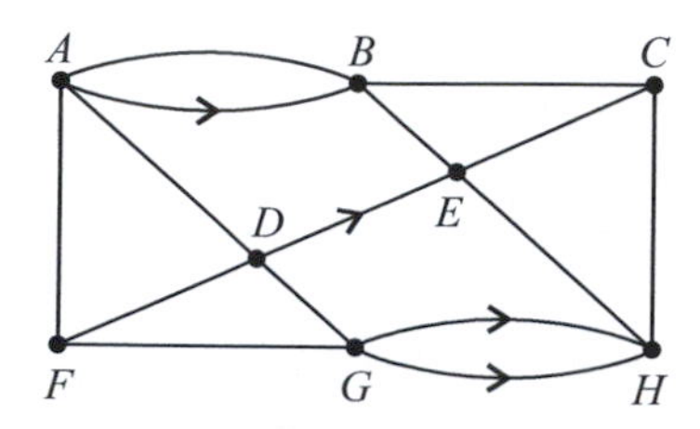

F23 In any graph that contains a closed Euler path, it is possible to assign directions to the edges in such a way that at each vertex, the number of incoming edges equals the number of outgoing edges. Explain how this can be done using the Euler path.

F24 We want to assign directions to the edges of this graph so that at each vertex, the number of incoming and outgoing edges are either equal or differ by 1.

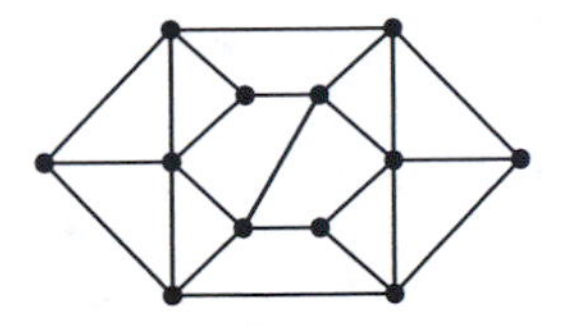

(a) Do this by the following strategy: Add edges joining the odd vertices in pairs and assign directions to the expanded graph by the method of the preceding problem. Finally, remove the added edges.

(b) Can a similar process be applied to any graph? Explain.

F25 Put in the edges, with labels, in this 8-vertex state diagram.

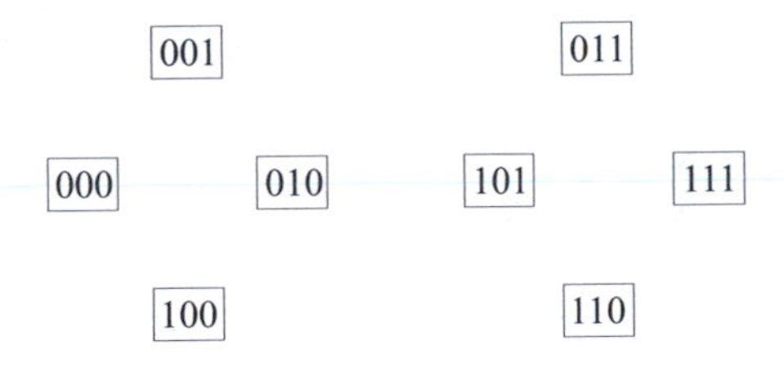

F26 Find an Euler path in the diagram above and use it to construct a DeBruijn Sequence of length 16 based on $\{0, 1\}$.

F27 (a) Describe how you would construct a DeBruijn sequence of length 32 based on $\{0, 1\}$ by using an appropriate state diagram.

(b) Explain how you know that there exists a DeBruijn sequence of length 2^n based on $\{0, 1\}$, for each positive integer n.

F28 Put in the edges, with labels, in this 9-vertex state diagram.

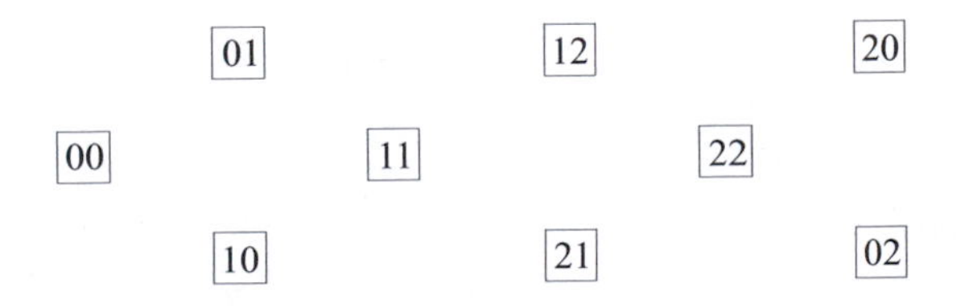

F29 Construct a DeBruijn sequence of length 27 based on the set of digits $\{0, 1, 2\}$. First guess what this probably means.

Hamilton Paths and Cycles

Definitions A *Hamilton path* in a graph **G** is a simple path that contains every vertex of **G**. A Hamilton cycle in **G** is a cycle that contains every vertex of **G**.

G1 In each graph below, find a Hamilton path or a Hamilton cycle, or both, if they exist.

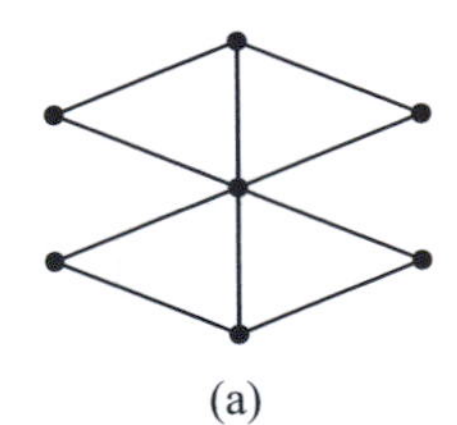

(a)

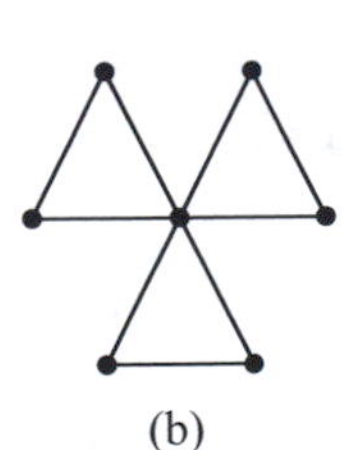

(b)

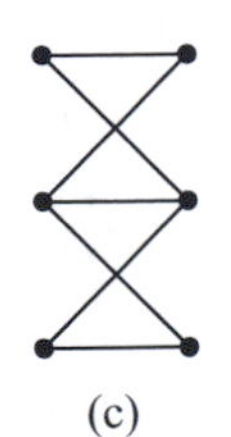

(c)

(d)

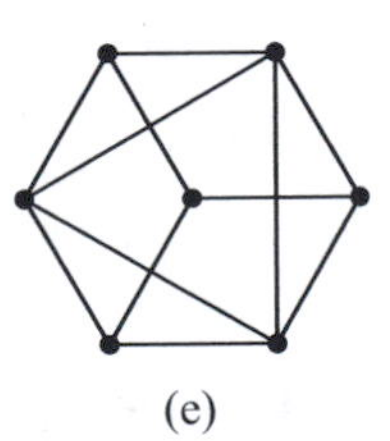

(e)

G2 Can a graph contain a Hamilton cycle but not a Hamilton path? Explain.

G3 Can a disconnected graph contain a Hamilton path or cycle?

G4 True or false? Every complete graph contains a Hamilton cycle.

Unlike the situation for Euler paths, there is no known test, or list of conditions (as in the Euler Path Theorem), that can be used to determine whether a given graph contains a Hamilton path or cycle. Instead, there are some negative tests, which can show that a given graph contains no Hamilton path or cycle, and some positive tests, which can show that a given graph contains a Hamilton path or cycle without actually constructing one. These tests, however, do not apply to every graph.

Some Negative Tests

We begin with a criterion that applies only to bipartite graphs.

Negative test for bipartite graphs

Suppose that **G** is a bipartite graph with m vertices of one color and n vertices of the other color. (As usual, each edge has one endpoint of each color.) Then:

(a) If $m \neq n$, **G** contains no Hamilton cycle.

(b) If m and n differ by 2 or more, then **G** contains no Hamilton path.

G5 Do either of these graphs contain Hamilton paths? Hamilton cycles?

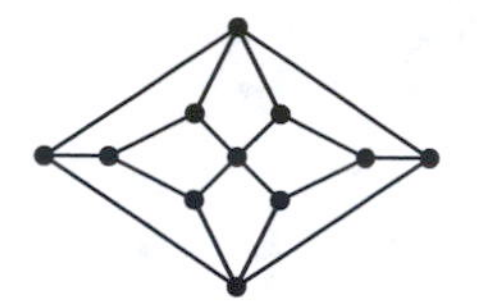

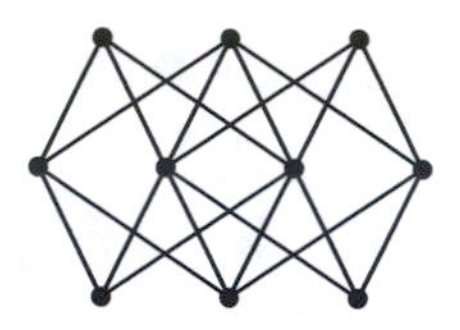

G6 Explain why this negative test is valid. (Suggestion: Think about the colors of the vertices along a Hamilton path or cycle.)

G7 If a bipartite graph contains an equal number of vertices of each color, does it necessarily contain a Hamilton cycle?

Next, we lead up to a more general negative test for Hamilton paths and cycles.

G8 (a) Suppose you cut a string at k points. How many pieces do you get?

(b) Answer the same question for a loop, such as a rubber band.

Now we consider removing some vertices from a graph. As usual, when a vertex is removed from a graph, all edges at that vertex are also removed.

G9 (a) If a graph consists of a simple path and k vertices are removed, how many components can the resulting subgraph contain? Indicate the possibilities.

(b) Answer the same question if k vertices are removed from a cycle graph.

G10 In the graph below, $ABCDEFGHIJK$ is a Hamilton path. For each given vertex set, find the number of components in the subgraph that results when the given set is removed from this graph.

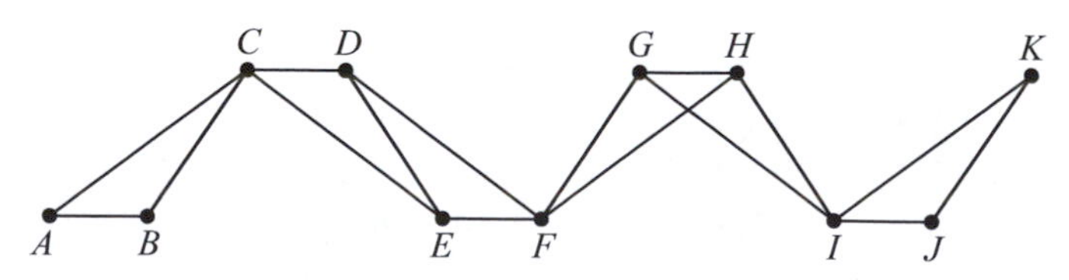

(a) $\{B, E, H\}$

(b) $\{C, F, I\}$

G11 In the graph below, $ABCDEFGHIJA$ is a Hamilton cycle. For each given vertex set, find the number of components in the subgraph that results when the given set is removed from this graph.

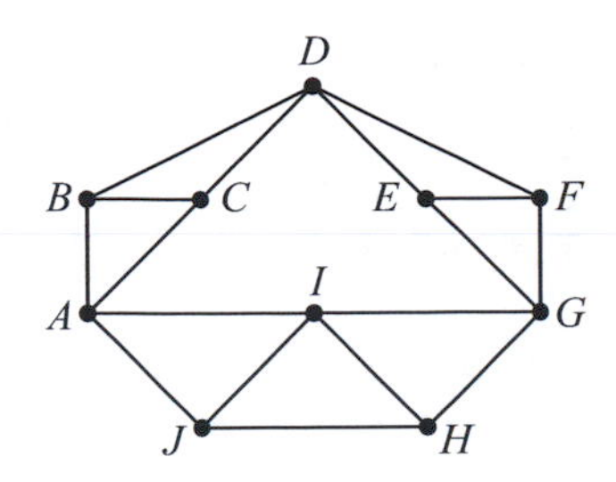

(a) $\{C, G, J\}$

(b) $\{A, D, G\}$

G12 Fill in the blanks, indicating the possibilities in each case.

(a) If k vertices are removed from a graph that contains a Hamilton path, then the number of components in the resulting subgraph is _______.

(b) If k vertices are removed from a graph that contains a Hamilton cycle, then the number of components in the resulting subgraph is _______.

Based on problem G12, we can say this:

Subgraph Test for Hamilton paths and cycles

1. Suppose that a graph **G** contains a set of k vertices (for some positive integer k) such that when these vertices are removed, the resulting subgraph contains at least $k + 2$ components. Then **G** contains no Hamilton path.

2. Suppose that a graph **G** contains a set of k vertices such that when these vertices are removed, the resulting subgraph contains at least $k + 1$ components. Then **G** contains no Hamilton cycle.

G13 Explain how the Subgraph Test follows logically from problem G12.

G14 Apply the Subgraph Test to each graph below. In each case, what is the conclusion?

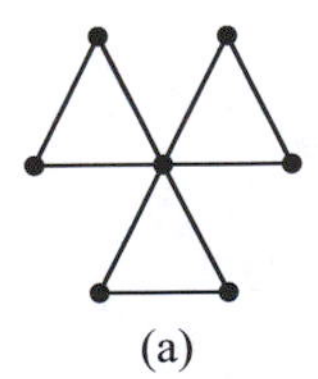
(a)

(b)

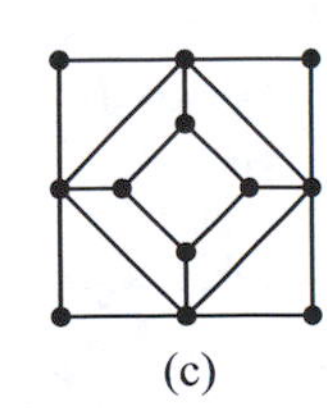
(c)

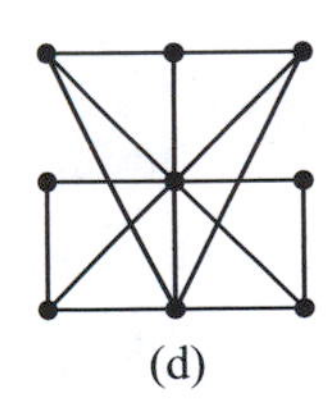
(d)

Positive Tests for Hamilton Cycles

Most of the research that has been done on Hamilton cycles is directed toward finding positive tests. (A positive test for Hamilton paths is given in problem G46.) We begin with the first, and probably best-known, result of this type.

Dirac's Theorem *Let* **G** *be a graph with n vertices, where* $n \geq 3$*, and suppose that each vertex has degree greater than or equal to* $n/2$*. Then* **G** *contains a Hamilton cycle.*

G15 According to Dirac's Theorem, this graph contains a Hamilton cycle. Find one.

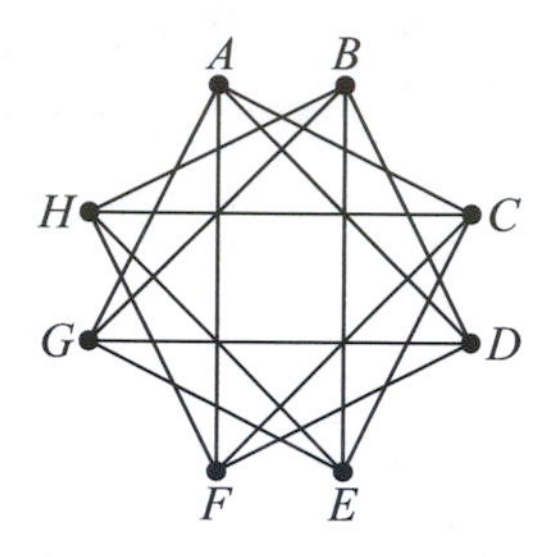

Next we give a positive test that is more useful than Dirac's Theorem because it applies to many more cases. In order to describe this, we begin by arranging the degree sequence of a given graph in increasing order:

$$d_1 \leq d_2 \leq d_3 \leq \cdots \leq d_n$$

In other words, d_1 is the smallest degree, d_2 is next, etc.

Posa's Theorem *Let* **G** *be a graph with n vertices, where* $n \geq 3$*, and (using the notation above) suppose that the degrees in* **G** *satisfy*

> ***Posa's degree conditions*** $d_1 > 1$, $d_2 > 2, \ldots,$ *continuing with* $d_i > i$*, for all values of* $i < n/2$.

Then **G** *contains a Hamilton cycle.*

This test is easy to apply. For example, in this graph with seven vertices, the degree sequence is $(3, 3, 4, 4, 4, 4, 4)$.

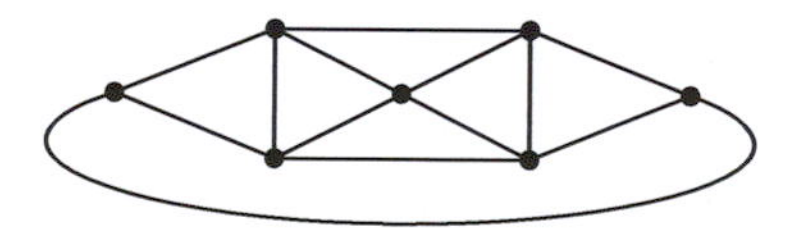

The condition $d_i > i$ must hold for all $i < 7/2$, which means $i = 1, 2,$ and 3. We easily check that $d_1 > 1$, $d_2 > 2$, and $d_3 > 3$. So Posa's Theorem tells us that the graph contains a Hamilton cycle. (Find it; it's an obvious one.)

G16 Which of the following degree sequences satisfy Posa's degree conditions?

(a) $(2, 3, 3, 3, 3, 3, 3)$
(b) $(2, 3, 3, 3, 3, 4)$
(c) $(3, 3, 4, 4, 4, 4, 4, 4)$
(d) $(3, 3, 4, 4, 4, 4, 4, 5, 5)$
(e) $(3, 3, 4, 5, 5, 5, 5, 5, 5)$
(f) $(2, 3, 4, 5, 5, 5, 5, 5, 5, 5)$

The next positive test for a Hamilton cycle is a little different. Instead of depending on the degree sequence of the graph, this test requires finding a Hamilton path whose endpoints satisfy a certain condition. In practice this is not always convenient to apply, but it is important because it provides the basis for proofs of all of the other positive tests.

The Path/Cycle Principle

Let **G** be a graph with n vertices, where $n \geq 3$, and suppose that **G** contains a Hamilton path whose endpoints (call them A and B) satisfy the condition $d(A) + d(B) \geq n$. Then **G** contains a Hamilton cycle.

Of course, if A and B are adjacent vertices in **G**, then the Hamilton path can obviously be extended to a Hamilton cycle. That's not very interesting. The Path/Cycle Principle becomes useful in cases like this graph in which there is an obvious Hamilton path whose endpoints are not adjacent, but which satisfy the necessary degree condition: $d(A) = 4$, $d(B) = 4$, and $n = 8$.

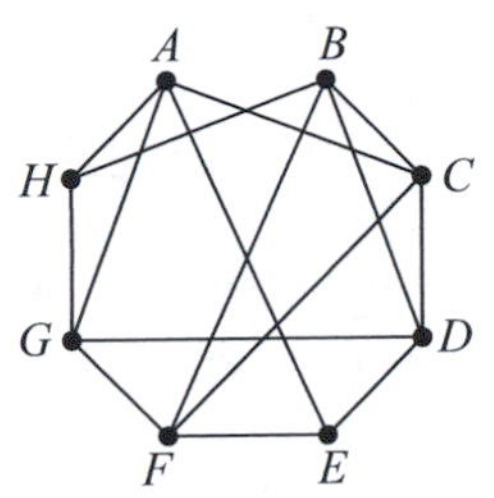

G17 Find a Hamilton cycle in the graph above.

G18 Apply the Path/Cycle Principle to the graph in problem G1(e).

Our final positive test for a Hamilton cycle is closely related to the Path/Cycle Principle, but it applies in many more cases. It allows us to expand a given graph by adding edges until we obtain a larger graph containing a Hamilton cycle. The test tells us that if the edges are added properly, then it is possible to conclude that the original graph also contains a Hamilton cycle.

The Bondy–Chvatal Theorem *Let* **G** *be a graph with n vertices, where* $n \geq 3$, *and suppose that* **G** *can be expanded to a larger graph* **G′** *that contains a Hamilton cycle, by adding edges one at a time in such a way that the following condition is satisfied*

> ***The Bondy–Chvatal condition*** An edge can be added joining vertices A and B if A and B are not already adjacent and if their degrees satisfy $d(A) + d(B) \geq n$.

Then **G** *contains a Hamilton cycle.*

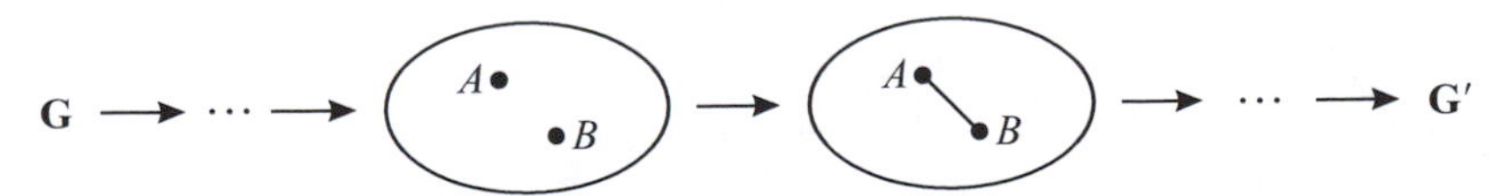

Note: The degrees $d(A)$ and $d(B)$ in the condition refer to the degrees of these vertices just before edge AB is added to the graph.

G19 Apply this to the graph preceding problem G17. Explain how the Bondy–Chvatal Theorem can be considered an extension of the Path/Cycle Principle.

G20 Show how graph **G** below can be expanded to **G′** by four steps satisfying the Bondy–Chvatal condition.

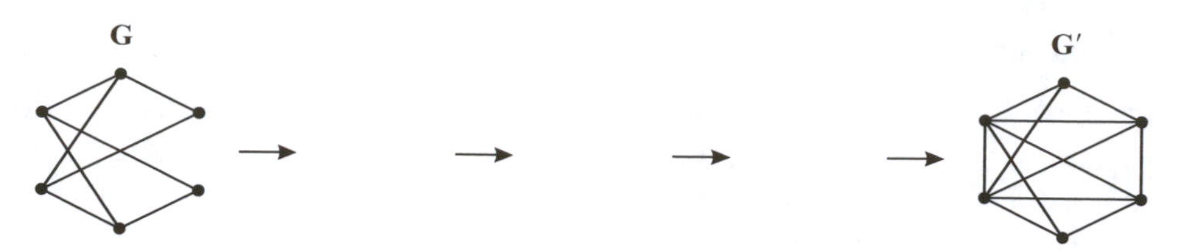

G21 (a) Apply the Bondy–Chvatal Theorem to this graph.

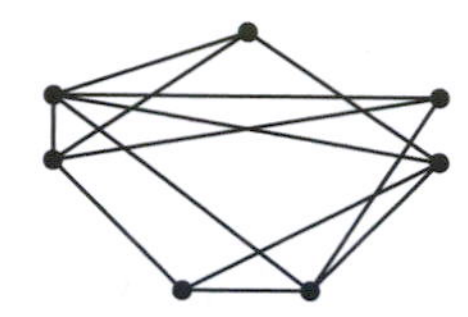

(b) Show how only one step of the Bondy–Chvatal Theorem is needed for the graph in (a) if we consider the degree sequence and apply Posa's Theorem.

(c) Find a Hamilton cycle in this graph.

Some Proofs

We begin by proving the Path/Cycle Principle and use it to prove the Bondy–Chvatal Theorem. Dirac's Theorem will then follow as an easy consequence of the Bondy–Chvatal Theorem. Posa's Theorem will be proved in problems G53–56.

Proof of the Path/Cycle Principle

We begin by labeling the vertices along a Hamilton path in our given graph **G**, as follows:

$$A \quad V_1 \quad V_2 \quad \cdots \quad V_{n-3} \quad V_{n-2} \quad B$$

The graph **G** can contain other edges not shown here, but it contains no other vertices. Remember that $d(A) + d(B)$ is greater than or equal to n; this will be important later. We will show that **G** contains a Hamilton cycle. For the purpose of this proof, it will be convenient to bend the diagram of the Hamilton path as shown below, so that it becomes easier to draw in other edges of the graph.

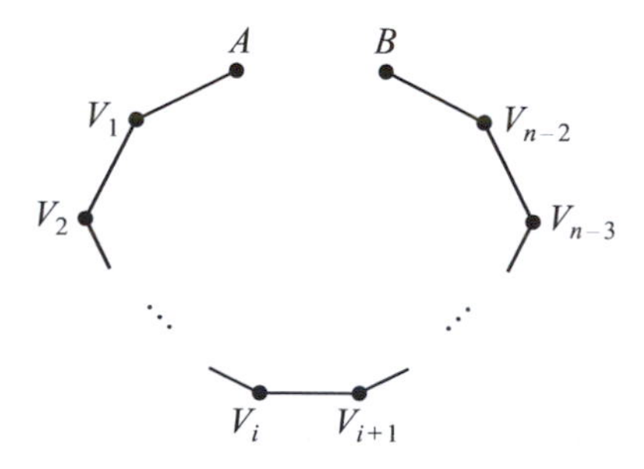

Notice that if **G** contains an edge from A to B, then obviously **G** contains a Hamilton cycle. So from here on we will restrict our attention to the only interesting case, in which A and B are nonadjacent vertices.

The idea of the proof is to show that there must exist some number i, where $1 \leq i \leq n-3$, such that **G** contains the two edges AV_{i+1} and BV_i shown in the diagram below. For example, these extra edges can be AV_2 and BV_1, or AV_3 and BV_2, etc.

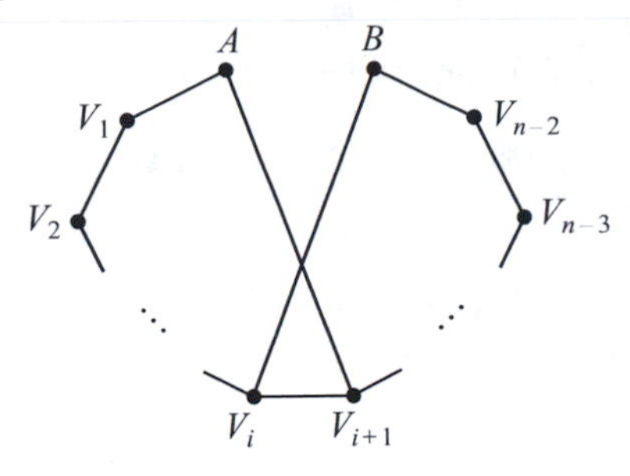

G22 If we can do that, how will it prove that **G** contains a Hamilton cycle?

G23 In the example below, which i-values, if any, satisfy the required condition? Here $n = 8$, so $1 \leq i \leq 5$. What Hamilton cycles result in this way?

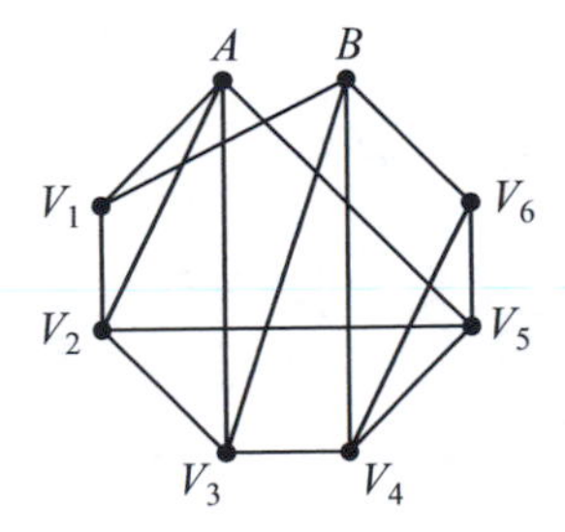

G24 For the graph in problem **G23**, define two sets of numbers S and T in the following way:

1. List all of the vertices that are adjacent to A, except for V_1. Subtract 1 from each subscript on the vertices in your list. The resulting numbers are the members of S.
2. List all of the vertices that are adjacent to B, except for V_6. The subscripts on these vertices are the members of T.

What is the significance of these two sets, in terms of the i-values that were found in problem **G23**?

Now we return to the general proof, using this diagram.

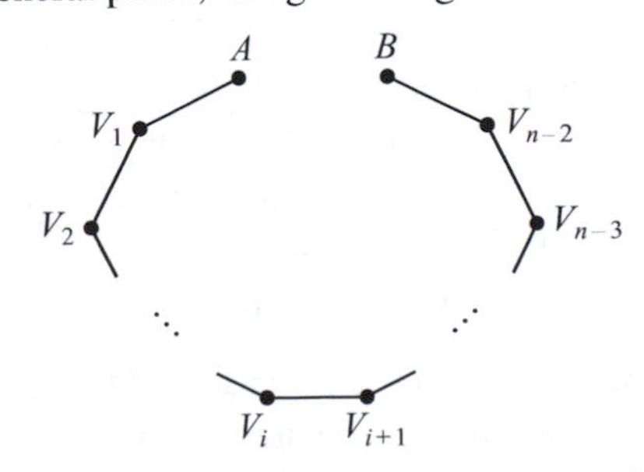

Remember that we have to show the edges AV_{i+1} and BV_i exist in **G** for some i-value such that $1 \leq i \leq n-3$. To do this we define two sets of numbers

$$S = \{\text{All } i \text{ such that } 1 \leq i \leq n-3 \text{ and } \_\_\_\_\_\_ \}$$
$$T = \{\text{All } i \text{ such that } 1 \leq i \leq n-3 \text{ and } \_\_\_\_\_\_ \}$$

G25 Complete the definitions of S and T and indicate what we have to prove about these sets.

G26 Explain how we know that the number of elements in S is $d(A) - 1$ and that the number of elements in T is $d(B) - 1$.

G27 Suppose you have two subsets of $\{1, 2, 3, \ldots, 19\}$, and each subset contains 10 numbers. Then what must be true about the two sets? Is the same conclusion valid if one set contains 9 numbers and the other contains 10? 9 and 11? 8 and 12?

G28 Returning to the two sets S and T defined in problem G27, show that the number of elements in S plus the number of elements in T is at least $n - 2$.

G29 What have we proved?

That completes the proof of the Path/Cycle Principle. Next we will use it to prove the Bondy–Chvatal Theorem.

Proof of the Bondy–Chvatal Theorem

Recall the situation: **G** has three or more vertices and can be expanded to a larger graph **G**$'$ by adding edges in such a way that each time a new edge AB is added to the graph, $d(A) + d(B) \geq n$ before the edge is added. Assuming that **G**$'$ contains a Hamilton cycle, we have to show that **G** also contains a Hamilton cycle.

This is done by contradiction: Suppose that

(∗) **G** does *not* contain a Hamilton cycle.

We will show that this is impossible by obtaining a contradiction. Look at the sequence of graphs that are obtained in the process of expanding from **G** to **G**$'$:

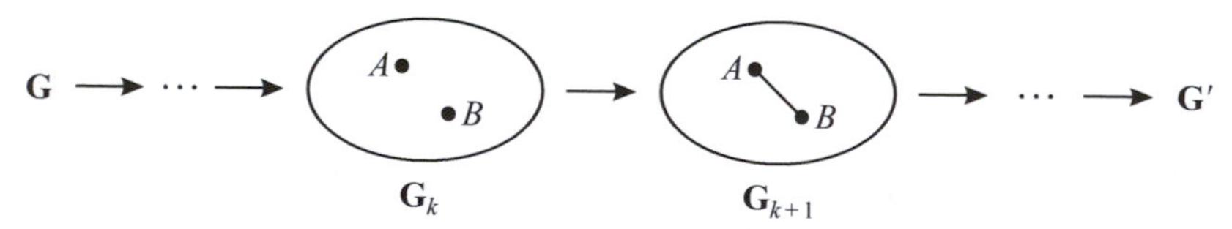

Eventually we reach **G**$'$, which contains a Hamilton cycle. But at the beginning, the graph $\mathbf{G}_1$ (which is **G**) does not contain a Hamilton cycle (we assumed this in (∗)), so there must be a graph $\mathbf{G}_k$ somewhere along the list which is the last one that contains no Hamilton cycle. After $\mathbf{G}_k$, all graphs on the list contain Hamilton cycles. We now focus our attention only on the graph $\mathbf{G}_k$ and the next one, $\mathbf{G}_{k+1}$. Let AB be the edge that gets added to the graph on the step from $\mathbf{G}_k$ to $\mathbf{G}_{k+1}$.

G30 Show that AB must be one of the edges in a Hamilton cycle in $\mathbf{G}_{k+1}$. (What impossible situation would occur if this were not true?)

G31 What do we know about $d(A) + d(B)$ in the graph $\mathbf{G}_k$?

G32 Apply the Path/Cycle Principle to $\mathbf{G}_k$. What does this show? But isn't that impossible?

We have obtained a contradiction. Therefore the assumption $(*)$ must be wrong. That proves **G** contains a Hamilton cycle.

Proof of Dirac's Theorem

This is now very easy, using the result of the Bondy–Chvatal Theorem.

G33 Starting with a graph that contains three or more vertices, each of which has degree greater than or equal to $n/2$, add an edge to **G** joining any two nonadjacent vertices. Then do it again. Continue doing this as long as possible. What graph do you eventually end up with?

G34 Apply the Bondy–Chvatal Theorem to show that **G** contains a Hamilton cycle. Be sure you know why the condition $d(A) + d(B) \geq n$ is satisfied on each step.

More Problems

G35 Find a Hamilton cycle in each graph below.

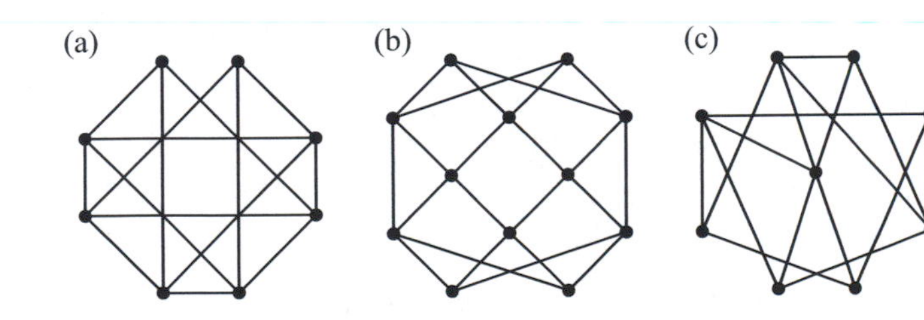

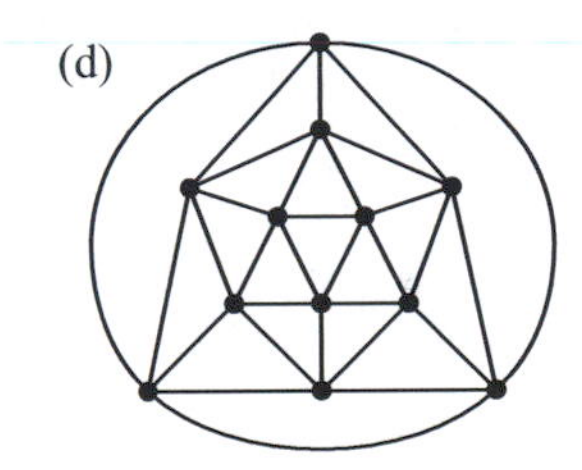

G36 Let m and n be positive integers. What must be true about m and n so that

(a) $\mathbf{K}_{m,n}$ contains a Hamilton cycle?

(b) $\mathbf{K}_{m,n}$ contains a Hamilton path?

G37 Apply the Subgraph Test to each graph below. In each case, what is the conclusion?

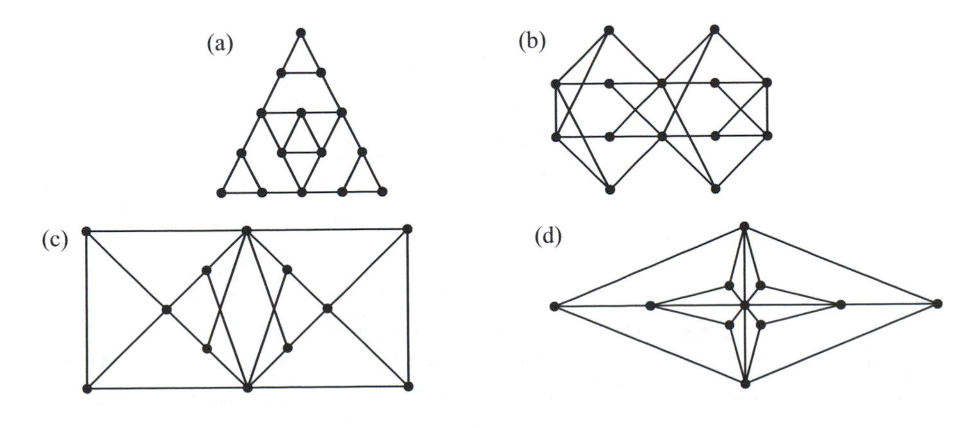

G38 Explain how the negative test for bipartite graphs can be regarded as a special case of the Subgraph Test.

G39 Which of the degree sequences below satisfy Posa's degree conditions?

(a) $(2, 2, 3, 4, 4, 5)$ (c) $(4, 5, 5, 5, 5, 6, 6, 6, 6, 6)$

(b) $(2, 3, 3, 4, 4, 5, 5)$ (d) $(4, 5, 5, 5, 5, 6, 6, 6, 6, 6, 6)$

G40 Show that if a graph satisfies Dirac's condition that each vertex has degree greater than or equal to $n/2$, then it also satisfies Posa's degree conditions.

G41 Apply the Path/Cycle Principle to this graph to show that there is a Hamilton cycle. Then find one.

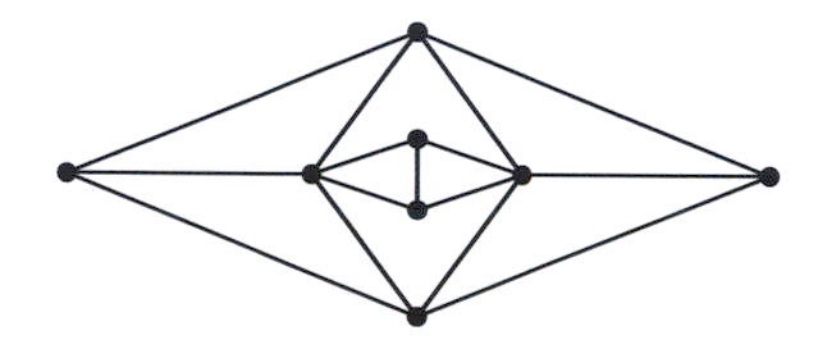

G42 Prove that every graph that has degree sequence $(2, 3, 3, 4, 4, 5, 5)$ contains a Hamilton cycle, even though this sequence doesn't satisfy Posa's degree conditions. Do this by the following steps:

(a) Show that if A is a vertex of degree 3, then there is a vertex B of degree 4 or 5 that is nonadjacent to A.

(b) What happens when an edge is added joining A and B?

G43 Apply the method of the preceding problem to the degree sequence $(4, 5, 5, 5, 5, 6, 6, 6, 6, 6, 6)$.

G44 Prove that every graph that has either degree sequence below contains a Hamilton cycle.

(a) $(2, 3, 3, 4, 5, 6, 7, 7, 7)$

(b) $(2, 3, 3, 4, 4, 6, 7, 7, 7, 7)$

G45 Use a combination of the Bondy-Chvatal Theorem and Posa's Theorem to show that this graph contains a Hamilton cycle.

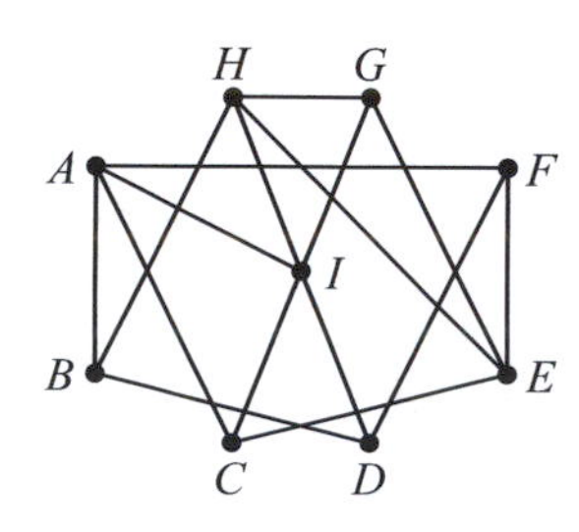

G46 Suppose that an edge is added to a graph **G** joining two vertices A and B that are not adjacent in **G**, and a Hamilton cycle exists in the expanded graph. Then what necessarily exists in **G**? (Note that we are not assuming here that $d(A) + d(B) \geq n$.)

G47 (a) Find a graph that has the degree sequence $(3, 3, 3, 3, 4, 4, 4)$ and contains no Hamilton cycle.

(b) Prove that every graph with the degree sequence above must contain a Hamilton path. (Hint: Add an edge to the graph so that the expanded graph satisfies Posa's degree conditions. Are you sure this is possible?)

G48 Find an example of a graph with three or more vertices that is regular, connected, bipartite, and contains no Hamilton cycle. (Suggestion: Try to arrange things so that the Subgraph Test applies.)

G49 For each graph below, find the sets S and T that are defined in the proof of the Path/Cycle Principle, and use them as in the proof to construct as many Hamilton cycles as possible.

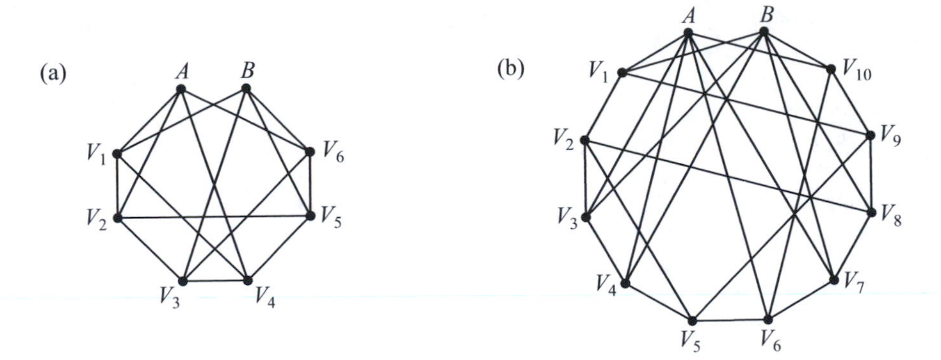

Ore's Theorem *Let* **G** *be a graph with n vertices, where* $n \geq 3$*, and suppose that* **G** *satisfies*

Ore's adjacency condition *Whenever* $d(A) + d(B) < n$ *for two vertices A and B, A and B are adjacent.*

Then **G** *contains a Hamilton cycle.*

G50 Use the Bondy–Chvatal Theorem to prove Ore's Theorem.

G51 Find a graph that satisfies Posa's degree conditions but not Ore's adjacency condition.

G52 Prove that every graph that satisfies Ore's adjacency condition also satisfies Posa's degree conditions by the following steps. Assuming that a graph satisfies Ore's adjacency condition, arrange the vertices $A_1, \ldots, A_n$ with degrees in increasing order $d_1 \leq d_2 \leq d_3 \leq \cdots \leq d_n$. For any $i < n/2$, let U denote the set of vertices $\{A_1, \ldots, A_i\}$ and let V denote the set of vertices $\{A_{i+1}, \ldots, A_n\}$. Suppose that $d_i \leq i$. We will obtain a contradiction.

(a) All vertices in U must be adjacent to each other. Why?

(b) Each vertex in U is adjacent to at most one vertex in V.

(c) V contains at least one vertex B that is not adjacent to any vertex in U. (Hint: Which set contains more vertices?)

(d) Show that $d(B) < n - i$.

(e) Find the contradiction.

Proof of Posa's Theorem

We will show that if **G** has at least three vertices, satisfies Posa's degree conditions, and is not a complete graph, then an edge can be added joining nonadjacent vertices A and B with degrees satisfying $d(A) + d(B) \geq n$. Continuing this way eventually results in the complete graph $\mathbf{K}_n$. It then follows by the Bondy–Chvatal Theorem that **G** contains a Hamilton cycle.

The key to proving the existence of suitable vertices A and B is to look at the sums $d_i + d_{i+1}$, where as usual the degrees are arranged in increasing order $d_1 \leq d_2 \leq d_3 \leq \cdots \leq d_n$.

G53 Show that if $d_1 + d_2 \geq n$, then A and B exist with the required properties. (Hint: Let A have degree d_1.)

In view of this result, we assume now that $d_1 + d_2 < n$ and proceed to consider sums $d_i + d_{i+1}$ for $i \geq 2$.

G54 Let r be the greatest integer less than $n/2$: $r = n/2 - 1$ if n is even or $(n-1)/2$ if n is odd. Show that $d_r + d_{r+1} \geq n$.

Along with the assumption that $d_1 + d_2 < n$, problem G54 implies that there must be some i in the range $2 \leq i \leq r$ such that $d_{i-1} + d_i < n$ and $d_i + d_{i+1} \geq n$.

G55 With i as above, show that $d_i < n - i$.

G56 Complete the proof of Posa's Theorem by showing that A and B exist with the required properties. (Hint: Let A have degree d_i.)

Planar Graphs

Definitions A *planar graph* is a graph that can be represented by a diagram in which no edges cross. Such a diagram is called a *plane diagram*. For example, $\mathbf{K}_4$ is a planar graph.

For example, two diagrams of $\mathbf{K}_4$ are shown below. The first is a plane diagram, while the second is not.

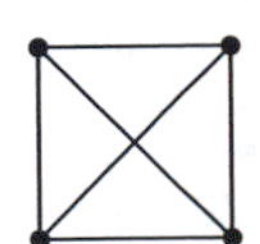

H1 Show that each graph below is planar by finding a plane diagram.

(a)

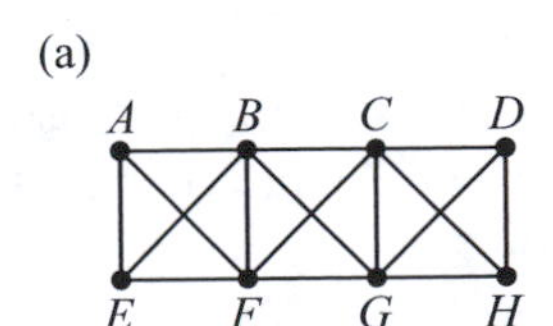

(b)

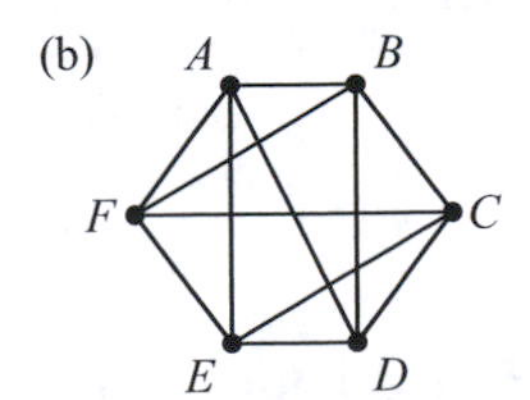

(c)

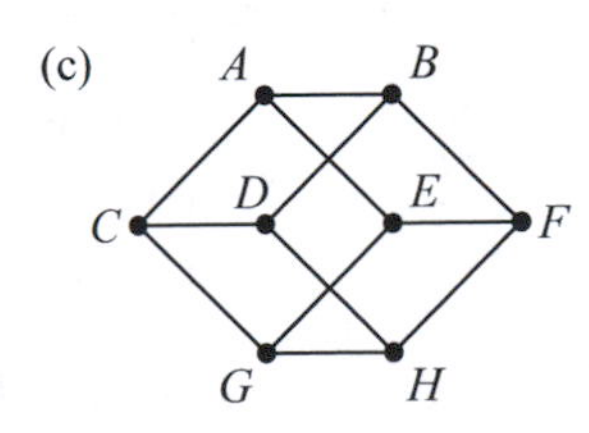

H2 If two graphs are isomorphic and one of them is planar, then is the other one necessarily also planar?

H3 Is every tree a planar graph? Every forest?

A reasonable approach to finding a plane diagram of a given graph is to start with the largest cycle you can find in the given graph and then try to draw in the remaining vertices and edges.

H4 Try that on this graph: If possible, begin with a Hamilton cycle.

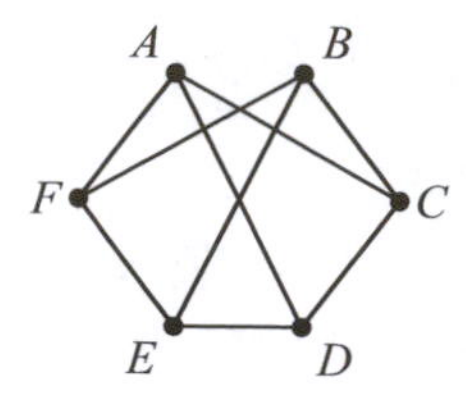

H5 What happens when the same procedure is applied to $\mathbf{K}_{3,3}$

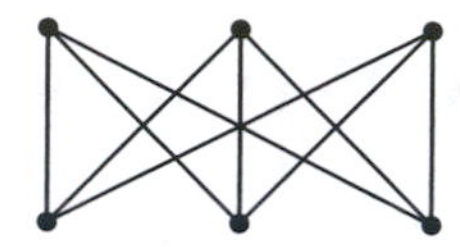

Since the attempt in problem H5 to find a plane diagram for $\mathbf{K}_{3,3}$ was unsuccessful, does that mean $\mathbf{K}_{3,3}$ is not a planar graph? In other words, can we be certain that there is not some other way to construct a plane diagram, possibly by arranging the vertices differently? The answer is yes: there is no other way to do it. The explanation is that the longest cycle, which in this case is a Hamilton cycle, would divide the plane into two regions, the inside and outside, in any plane diagram. Each of the remaining three edges would have to go either inside or outside of the cycle, and it is easy to try all possibilities and verify that there is no way these edges can be drawn in without some edges crossing. By this reasoning we can conclude that $\mathbf{K}_{3,3}$ is a non-planar graph.

H6 Use the same method to show that $\mathbf{K}_5$ is a non-planar graph.

Later we will see another way to prove that $\mathbf{K}_{3,3}$ and $\mathbf{K}_5$ are non-planar, without relying on a trial-and-error approach. Knowing that these two graphs are non-planar will also help us to prove that other graphs are non-planar:

H7 Suppose that $\mathbf{G}$ is a graph that contains a subgraph which is isomorphic to either $\mathbf{K}_{3,3}$ or $\mathbf{K}_5$. Explain how you know that $\mathbf{G}$ must be non-planar.

We will see that in a certain sense, $\mathbf{K}_{3,3}$ and $\mathbf{K}_5$ are present inside every non-planar graph, although not necessarily as subgraphs; the situation is a little more complicated than that.

Regions Formed by a Plane Diagram

If $\mathbf{G}$ is a planar graph, then any plane diagram of $\mathbf{G}$ divides the plane into regions, one of which is the infinite outer region. For example, the first diagram below divides the plane into three regions. The same graph can also be represented by different plane diagrams, as shown here.

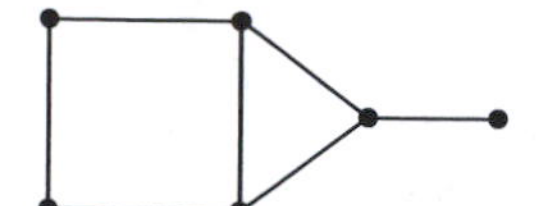 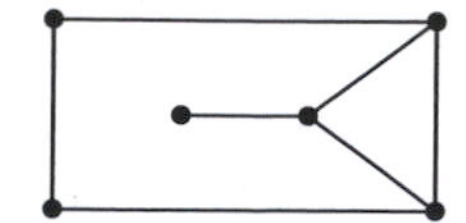 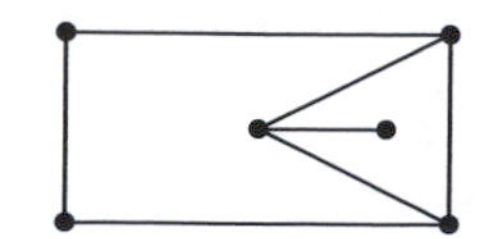

While each diagram divides the plane into three regions, notice that there are some differences. Specifically, look at the border, or boundary, of each region and imagine that you are walking around along a fence. When you return to your starting point, you have completed a closed path. In the first diagram of **G**, these closed paths are cycles for the two inner regions (a 4-cycle and a 3-cycle), and a more complicated closed path in the case of the outer region. We are interested in the lengths of these closed paths. These are 4, 3, and 7 in the case of this first diagram. (Be sure you agree that the boundary of the outer region has length 7.) These lengths are called the *regional degrees* of the plane diagram.

H8 Find the regional degrees in the two other diagrams shown above.

Notice that the regional degrees can vary from one diagram of a graph to another. This is not the case for the degrees of vertices. Thus the regional degrees, unlike vertex degrees, depend on the plane diagram of a graph and not just on the graph itself. However, some things remain unchanged: in all diagrams of **G** the number of regions is the same. (We will see below why this is true.) Also, notice something else:

H9 In each plane diagram on the previous page, add the regional degrees and compare the sum with the number of edges in the graph. Fill in the blank in the theorem below and explain why this is true.

The Regional Degree Theorem *Let* **G** *be a connected planar graph, and let* $r_1, r_2, r_3, \ldots$ *be the degrees of the regions in any plane diagram of* **G**. *Then the sum* $r_1 + r_2 + r_3 + \cdots$ *is equal to* ____.

Notice that this theorem is stated here only for planar graphs that are connected. The reason for this is to guarantee that the boundary of each region consists of a single closed path.

There is an important relationship between the number of vertices, edges, and regions in any plane diagram of a connected planar graph:

Euler's Formula *Let* **G** *be a connected planar graph with v vertices and e edges, and let r be the number of regions in any plane diagram of* **G**. *Then*

$$v - e + r = 2.$$

Try this on some familiar graphs. Notice that this explains why the number of regions is the same in all plane diagrams of a given graph, even though the degrees of the regions can differ from one diagram to another.

As an application of Euler's Formula, we will give a second proof of the fact that $\mathbf{K}_5$ is non-planar, this time by contradiction.

H10 If $\mathbf{K}_5$ were a planar graph, how many regions would a plane diagram of $\mathbf{K}_5$ have to contain?

Before continuing, we need one fact about regional degrees:

Let **G** *be any connected planar graph with three or more vertices. Then in any plane diagram of* **G**, *every region has degree greater than or equal to 3.*

H11 Why do we have to assume that **G** has three or more vertices? What are the regional degrees in graphs with one or two vertices?

Proof that $\mathbf{K}_5$ is Non-Planar, Using Euler's Formula

Suppose that $\mathbf{K}_5$ is a planar graph. We will show that this assumption leads to a contradiction.

H12 Show that in any plane diagram of $\mathbf{K}_5$, the sum of the regional degrees is greater than or equal to 21.

H13 On the other hand, this is impossible. Why?

This contradiction shows that our assumption that $\mathbf{K}_5$ is planar must be wrong.

Another application of Euler's Formula involves 3-dimensional solids, such as these:

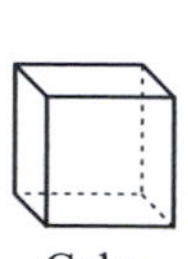
Cube

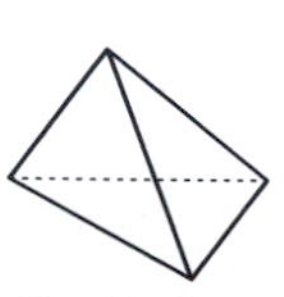
Tetrehedron

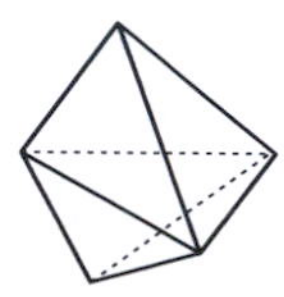
Hexahedron

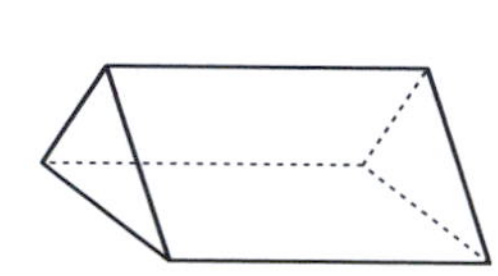
Triangular Prism

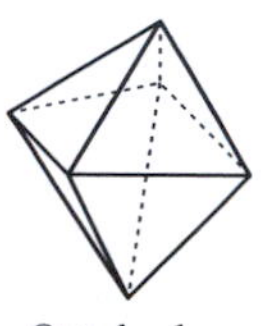
Octahedron

H14 For each figure shown above, count the number of vertices v (corners), edges e, and faces f (flat surfaces). Find $v - e + f$.

H15 Obviously the result in problem H14 is closely related to Euler's Formula. What exactly is the connection?

You may notice that we still haven't proved Euler's Formula. This is actually surprisingly easy, and we'll do it now. The idea is to reduce a given connected planar graph down to a spanning tree. On each step we remove one edge from a cycle, keeping all vertices. Eventually we will be left with a spanning tree.

We begin with a given plane diagram in which there are v vertices, e edges, and r regions.

H16 What happens to the numbers v, e, and r on each step? What happens to the value $v - e + r$?

H17 What is $v - e + r$ at the end of the reduction process, when the graph is reduced to a spanning tree? So what was it at the beginning?

That proves Euler's Formula for connected planar graphs, and consequently also the formula $v - e + f = 2$ for 3-dimensional solids. Notice the advantage of using graph theory to prove this: Working only with solid figures, there would be no way to reduce down to a tree.

H18 A dodecahedron has 12 faces and 20 vertices. Each face has the same number of edges. Use Euler's Formula and regional degrees to determine what that number must be.

H19 Repeat problem H18 for an icosahedron, which has 20 faces and 12 vertices.

Non-Planar Graphs and Kuratowski's Theorem

We saw in problem H7 that a graph can be proved to be non-planar by showing, if possible, that it contains a subgraph isomorphic to either $\mathbf{K}_5$ or $\mathbf{K}_{3,3}$. However, consider these graphs:

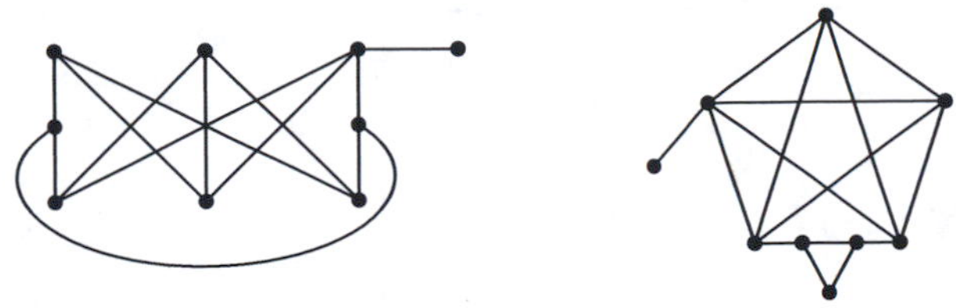

H20 In each case, find a subgraph that is "obviously" non-planar, based on what we know about $\mathbf{K}_5$ and $\mathbf{K}_{3,3}$. Is the subgraph isomorphic to either $\mathbf{K}_5$ or $\mathbf{K}_{3,3}$?

The problem above suggests the following idea: If a non-planar graph is changed by inserting vertices of degree 2 along an edge, converting the edge into a simple path, then the resulting graph is still non-planar. This process can be repeated any number of times on any number of edges, and the graph remains non-planar. On the other hand, if a graph is planar, then the insertion of these new vertices does not change that, either.

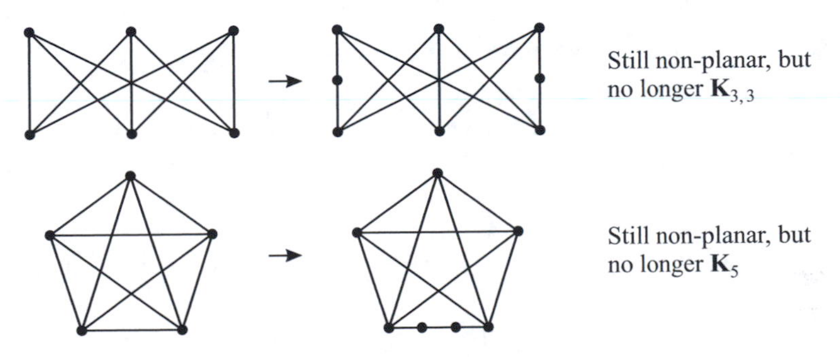

A graph that is obtained by inserting vertices of degree 2 in this way is called a *subdivision* of the original graph. The examples above are subdivisions of $\mathbf{K}_{3,3}$ and $\mathbf{K}_5$.

The important fact about subdivisions is this:

If one graph is a subdivision of another, then either both graphs are planar or else both are non-planar.

H21 Explain why this is true.

To clarify the definition of a subdivision, we emphasize that only vertices of degree 2 can be inserted if the new graph is to be a subdivision of the original one. For example, the insertion of a vertex shown below does not result in a subdivision of $\mathbf{K}_4$.

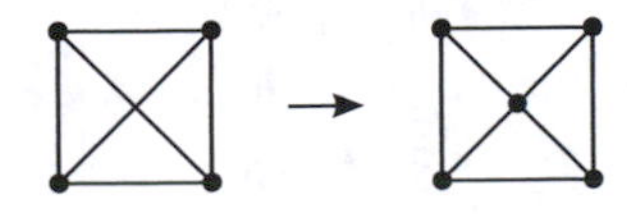

Putting things together, we can now say this:

If a graph **G** *contains a subgraph which is either* $\mathbf{K}_5$ *or* $\mathbf{K}_{3,3}$ *or a subdivision of one of these, then* **G** *must be non-planar.*

It is interesting to see how this observation applies in cases where it is not obvious what the subgraph should be. For example, the graph **G** below has subgraphs $\mathbf{H}_1$ and $\mathbf{H}_2$ as shown.

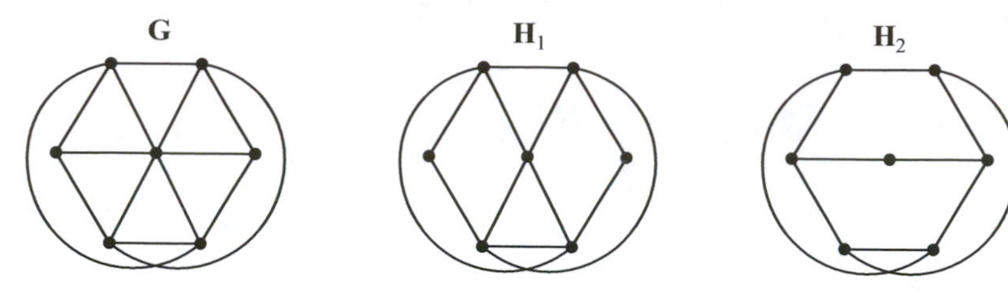

H22 Is $\mathbf{H}_1$ a subdivision of anything familiar? What about $\mathbf{H}_2$? What does this indicate about **G**?

Finding an appropriate subgraph, to show that a given graph is non-planar, is not always easy. How do we decide what to remove? This usually requires some trial and error, but there are some clues:

Clue #1 We want a non-planar subgraph, so don't remove too many edges that cross.

In the example **G** above, we would know to not remove either of the curved edges.

Clue #2 Any subdivision of $\mathbf{K}_5$ has the degree sequence $(4, 4, 4, 4, 4, 2, \ldots)$, where there can be any number of vertices of degree 2. Any subdivision of $\mathbf{K}_{3,3}$ has the degree sequence $(3, 3, 3, 3, 3, 3, 2, \ldots)$.

H23 In looking for an appropriate subgraph to show that this graph is non-planar

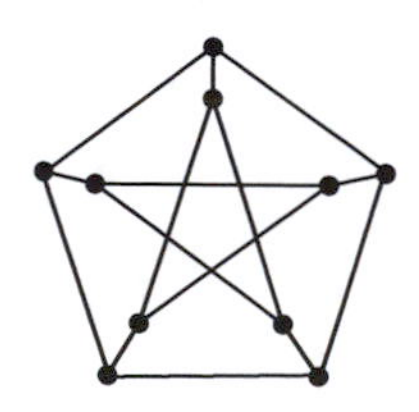

(a) Explain why you would not look for $\mathbf{K}_5$ or a subdivision of $\mathbf{K}_5$.

(b) Show that you wouldn't want to remove more than one edge from the center "star."

(c) Find $\mathbf{K}_{3,3}$ or a subdivision of $\mathbf{K}_{3,3}$ in this graph.

Finally, we note that every non-planar graph can be proved to be non-planar by finding an appropriate subgraph, although that's not always the easiest way to do it. This conclusion is based on the following theorem, which we state here without attempting to prove it.

Kuratowski's Theorem *Every non-planar graph contains a subgraph which is either* $\mathbf{K}_5$, $\mathbf{K}_{3,3}$, *or a subdivision of* $\mathbf{K}_5$ *or* $\mathbf{K}_{3,3}$.

More Problems

H24 Find a plane diagram for each graph below.

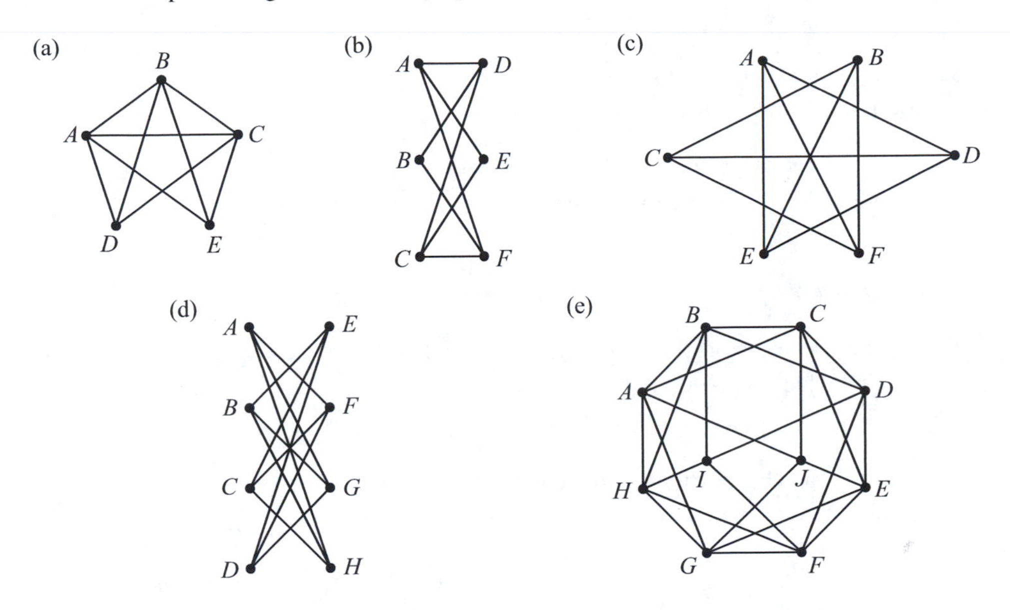

H25 Use a trial-and-error approach to prove that this graph is non-planar. Begin with a Hamilton cycle.

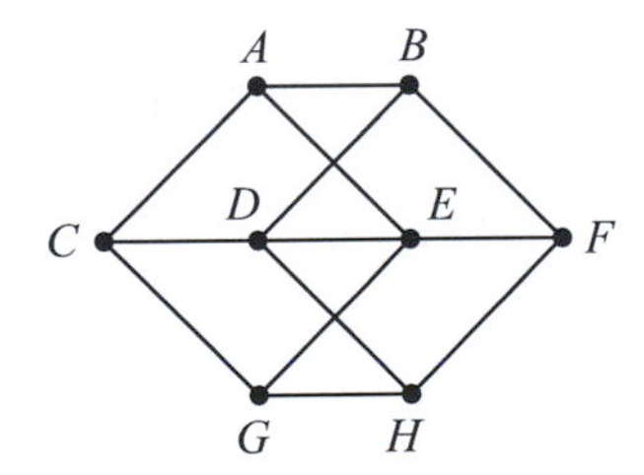

H26 Let r be the number of regions in a plane diagram of a disconnected planar graph **G**. Suppose that **G** has k components.

(a) Find the value of $v - e + r$ in terms of k if **G** is a forest.

(b) Find $v - e + r$ any planar graph with k components. (See the proof of Euler's Formula if necessary.)

H27 Let r be the number of regions in a plane diagram of a 3-regular connected planar graph. Let v and e be as usual. (a) Find e in terms of v. (b) Find r in terms of v.

H28 A soccer ball has 32 faces, each of which is a regular pentagon or hexagon. Because of the angles involved, exactly three faces meet at each corner. Without looking at a ball, determine how many of each type of face there are.

H29 Let **G** be a connected planar graph with three or more vertices, and let v, e, and r be as usual. (a) Use the regional degrees in any plane diagram of **G** to show that r is less than or equal to $2e/3$. (b) Prove that e is less than or equal to $3v - 6$.

H30 Use the result of problem H29 to prove (again) that $\mathbf{K}_5$ is non-planar.

H31 Use the result of problem H29 to prove that in any planar graph, there must be some vertex of degree less than or equal to 5.

H32 In any plane diagram of a bipartite planar graph, all of the regional degrees must be even. Explain why this is true.

H33 Suppose that the graph in problem H29 is bipartite. Show that r is less than or equal to $e/2$ and that e is less than or equal to $2v - 4$. Use this to give another proof that $\mathbf{K}_{3,3}$ is non-planar.

H34 Show that there cannot exist a 4-regular bipartite planar graph.

H35 (a) Find an example of a connected graph that has the degree sequence $(4, 4, 4, 4, 4, 2, \ldots)$, but which is not a subdivision of $\mathbf{K}_5$.

(b) Answer the same question for $(3, 3, 3, 3, 3, 3, 2, \ldots)$ and $\mathbf{K}_{3,3}$.

H36 In each graph below, find a subgraph that is either $\mathbf{K}_5$, $\mathbf{K}_{3,3}$, or a subdivision of one of these. For (e), find subdivisions of both.

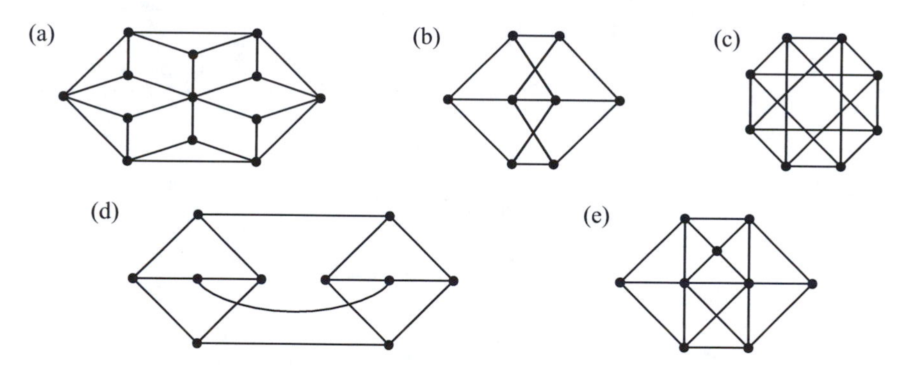

H37 (a) Show that this graph is non-planar by the trial-and-error method, starting with a Hamilton cycle.

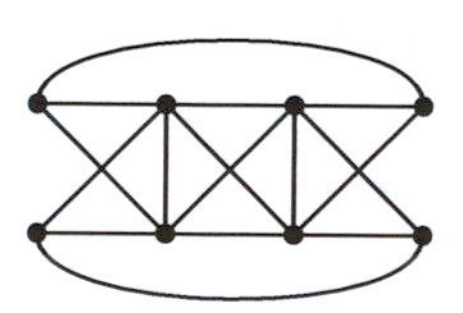

(b) Find either $\mathbf{K}_5$, $\mathbf{K}_{3,3}$, or a subdivision of one of them in this graph.

I

Independence and Covering

Definitions An *independent vertex set* in a graph is a set of vertices that does not include any adjacent vertices. An *independent edge set* is a set of edges that does not include any edges that share an endpoint.

For example, in this graph

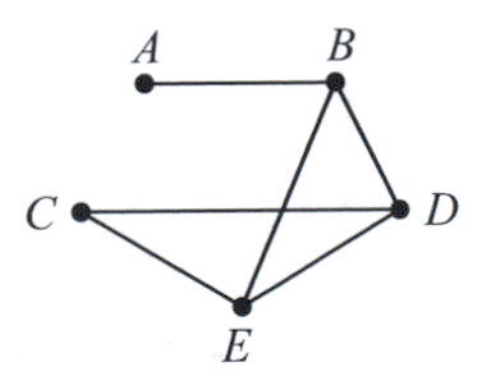

$\{A, C\}$ is an independent vertex set. $\{AB, CD\}$ is an independent edge set. Alternatively, in an unlabeled graph, these same sets can be indicated this way:

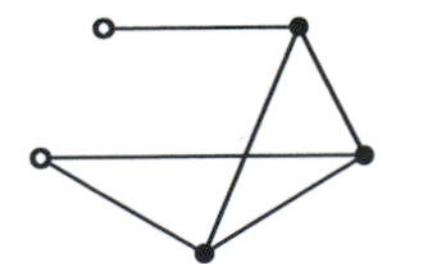

In any graph, a set consisting of just one vertex is always an independent vertex set since it contains no adjacent vertices. A set consisting of just one edge is an independent edge set.

The Independence Numbers of a Graph

We are interested in independent vertex sets and independent edge sets that are as large as possible in a given graph. These are called *maximal independent sets*. The *independence numbers* α and α' for a given graph are defined as

$$\alpha = \text{ the number of vertices in a maximal independent set}$$

$$\alpha' = \text{ the number of edges in a maximal independent edge set}$$

11 Find α and α', along with corresponding maximal independent vertex and edge sets, for each graph below.

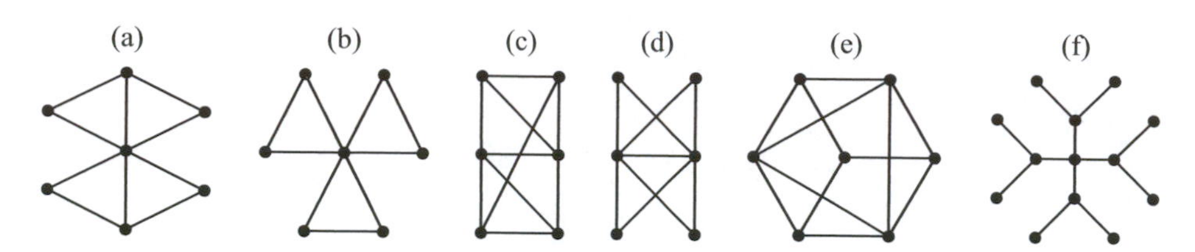

12 (a) In a graph with seven vertices, explain why α' could not be 4.

(b) Fill in the blank: In a graph with n vertices, α' is always less than or equal to _____.

(c) Find an example showing that α does not always satisfy this same inequality.

13 (a) Suppose that $\{A, B, C\}$ is an independent vertex set in a certain graph **G**, and suppose that the degrees of A, B, and C are 3, 4, and 5. Explain how you know that **G** must contain at least 12 edges.

(b) In any graph, what can you say about the sum of the degrees in a maximal independent vertex set, in relation to the number of edges? Explain why this is true.

14 (a) Use degrees to show that this graph does not contain five independent vertices. (Hint: Show that certain vertices would have to be included.)

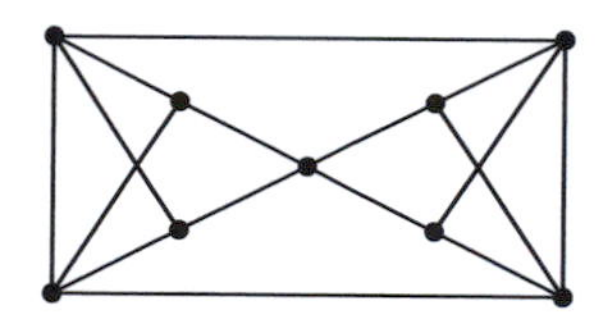

(b) What are the values of α and α' for this graph?

This summarizes what we have observed so far:

Independence Theorem #1 *In a graph with n vertices and e edges,*

(a) α' is less than or equal to $n/2$.

(b) If $d_1, d_2, \ldots$ are the degrees in a maximal independent vertex set, then the sum $d_1 + d_2 + \cdots$ is less than or equal to e.

Using (b), we can prove something that applies to regular graphs:

Corollary *In any d-regular graph with n vertices and $d \geq 1$, α is less than or equal to $n/2$.*

15 Show how this follows from the theorem by these steps:

(a) What is the product nd equal to in any regular graph?

(b) What can you say about the product αd?

(c) Combine (a) and (b) to prove the corollary.

Next we notice something about the sum $\alpha + \alpha'$ in any graph:

I6 For each graph in problem I1, compare the sum $\alpha + \alpha'$ with n, the number of vertices in the graph. What seems to always be true?

Independence Theorem #2 *In any graph with n vertices, $\alpha + \alpha'$ is always less than or equal to n.*

Before getting into the proof of this, we show how it can be used:

I7 In this graph,

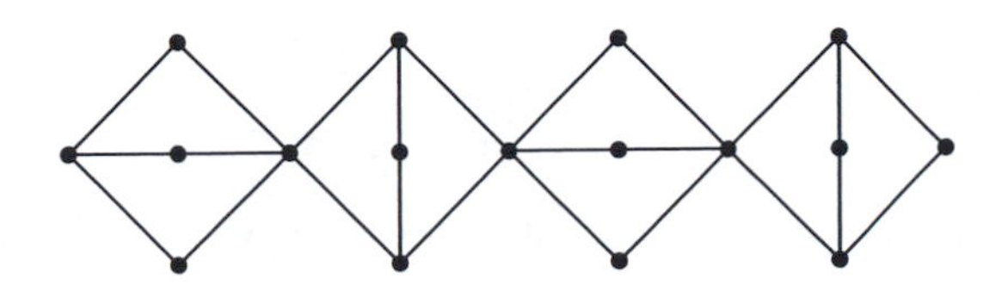

(a) Find 10 independent vertices and 7 independent edges.

(b) Use the theorem to show that these sets must be maximal without searching for larger independent sets.

Proof of Independence Theorem #2 In a graph **G**, let **S** be a maximal independent vertex set and let **S**$'$ be a maximal independent edge set. **S** contains α vertices and **S**$'$ contains α' edges. We will prove the theorem by constructing a set **X** consisting of $\alpha + \alpha'$ distinct vertices.

We begin by taking all vertices in **S**. These are included in **X**.

I8 Every edge in **G** has at least one endpoint that is not in **S**. Why?

Now, for each edge in the set **S**$'$, take an endpoint that is not in **S** and add that vertex to **X**.

I9 Show that by doing this, we add exactly α' new vertices to **X**. (How do you know that they are all different?)

I10 Explain why this proves the theorem.

The application of Independence Theorem #2 in problem I7 was possible only because in that graph, $\alpha + \alpha' = n$. This is not true in all graphs, but the next theorem shows that it is always true in bipartite graphs:

Independence Theorem #3 *In any bipartite graph with n vertices, $\alpha + \alpha' = n$.*

This is one form of what is usually called the König–Egervary Theorem, and is considerably harder to prove than the previous theorems in this chapter. The proof will appear in chapter M.

As a consequence of Independence Theorem #3, we have this, for regular bipartite graphs:

Corollary *In any d-regular bipartite graph with n vertices and $d \geq 1$, $\alpha = \alpha' = n/2$.*

I11 Prove this by using the theorem and results from earlier in this chapter.

When $\alpha' = n/2$, a maximal independent edge set has the additional property that it covers all of the vertices in the graph. This means that each vertex is an endpoint of some edge in the set. For example, in this graph, the independent edge set $\{AB, CD, EF, GH\}$ covers all of the vertices.

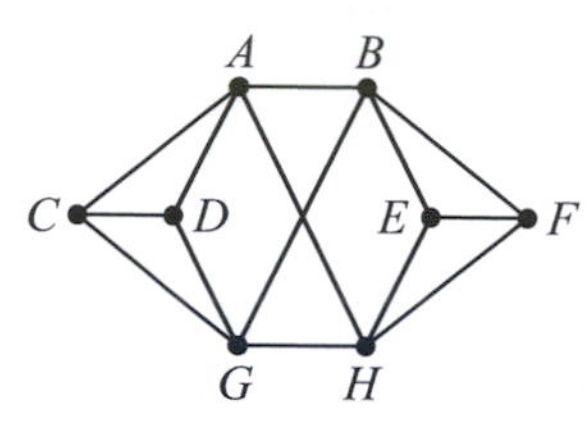

In general, an independent edge set that covers all of the vertices in a graph is called a *perfect matching*. The corollary to Independence Theorem #3 shows that every d-regular bipartite graph contains a perfect matching, as long as $d \geq 1$. (As the example above shows, other graphs not covered by the corollary can also contain perfect matchings.)

I12 Find a perfect matching in each graph below.

(a)

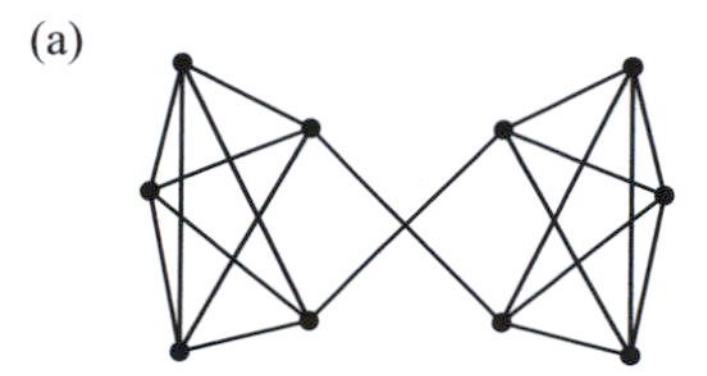

(b)

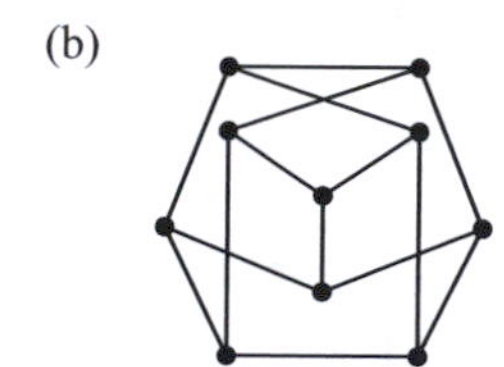

A Graph Game

Now that we know about perfect matchings, we can play a game using a graph. For any graph **G**, the game proceeds as follows: Two players take turns selecting vertices of **G**. The first player begins by selecting any vertex A, and then the second player must select a vertex B that is adjacent to A. The first player goes again, selecting a vertex C that is adjacent to B and not previously selected. In this way a simple path is formed. This continues until one of the players is unable to select a vertex. The last player to select a vertex is the winner.

Example (numbers indicate order in which vertices are selected)

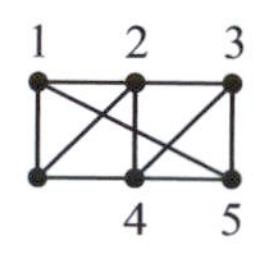

First player wins

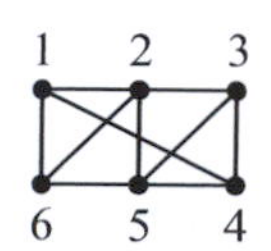

Second player wins

I13 In this graph, the second player has a winning strategy. Do you see what it is? Here is a generous hint.

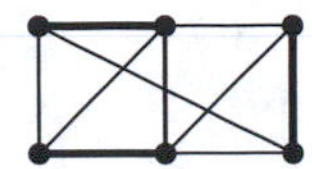

Surprisingly, no matter what graph is used, either the first or second player always has a winning strategy. That is, this player can be sure of winning by playing a certain way, no matter what the other player does. Without giving the secret away, we'll just say that it has something to do with whether or not a perfect matching exists in the graph.

I14 Try the game using this graph. This time you want to go first.

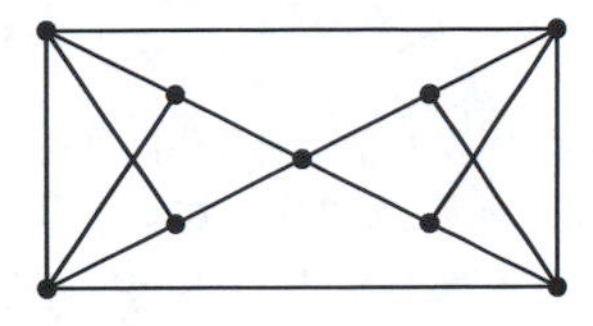

(Hint: Find four independent edges and start at the uncovered vertex.)

Covering Sets and Covering Numbers

A *covering edge set* in a graph is a set of edges with the property that every vertex in the graph is an endpoint of at least one edge in the set. (Note: A perfect matching is a covering edge set that is also independent. In general, a covering edge set doesn't have to be independent.)

For example, one covering edge set in this graph would be $\{AB, CD, CE\}$.

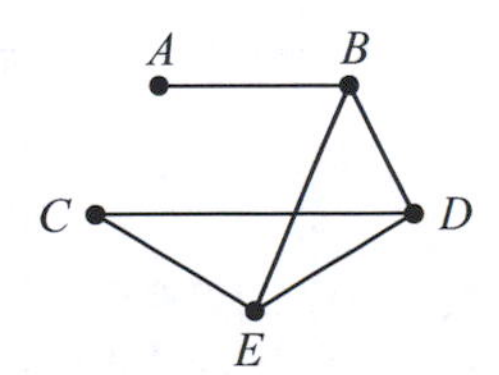

A *covering vertex set* in a graph is a set of vertices with the property that every edge in the graph has at least one endpoint in the set.

In the example above, one covering vertex set is $\{B, C, D\}$.

For covering sets, we are interested in the smallest possible ones in a given graph. These are called *minimal covering sets*.

I15 Show that the covering edge set and the covering vertex set given in the preceding example are each minimal.

The *covering numbers* for a given graph are

$$\beta = \text{the number of vertices in a minimal covering vertex set,}$$

$$\beta' = \text{the number of edges in a minimal covering edge set.}$$

We will see that these numbers are closely related to the independence numbers α and α'.

I16 For each of these graphs, find the values of β and β', along with corresponding minimal covering sets.

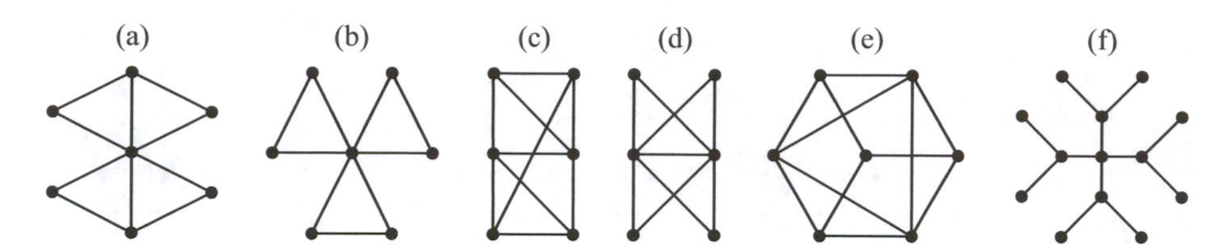

I17 Use your answers from above and from problem I1 to find the sums $\alpha+\beta$ and $\alpha'+\beta'$ in each graph. Guess what is always true. (Fill in below.)

In any graph with n vertices, $\alpha + \beta =$ ____ and $\alpha' + \beta' =$ ____.

Actually, we have to be a little more careful than that: β' does not make sense in a graph that contains any vertices of degree 0 (why not?). Therefore any time we mention β' we will have to assume that each vertex in the graph has degree greater than or equal to 1. Or equivalently, we can say that the minimum degree δ is greater than or equal to 1.

The relations above are explained by looking at how minimal covering sets can be constructed starting with maximal independent sets. The case for vertex sets is simpler, so we begin with that.

Covering Theorem #1 *In any graph, let* **S** *be a maximal independent vertex set. Then a minimal covering vertex set can be obtained by taking all vertices that are not in* **S**.

I18 Try this for each graph in problem I16 (a), (c), and (f).

An immediate consequence of Covering Theorem #1 is this:

Corollary *In any graph with n vertices,* $\alpha + \beta = n$.

Covering Theorem #1 will be proved in problems I34 and I35.

Now consider edge sets. We will see how a minimal covering edge set can be obtained from a maximal independent edge set:

Covering Theorem #2 *In any graph with minimum degree* $\delta \geq 1$, *let* **S**$'$ *be a maximal independent edge set. Then a minimal covering edge set can be obtained in the following way: Start by including all edges in* **S**$'$. *Then, for each vertex that is not covered by* **S**$'$, *select any one edge at that vertex and include that edge.*

I19 Try this for each graph in problem I16 (d), (e), and (f).

As for why this theorem is true, notice that it is obvious that the edge set constructed by the theorem is a covering edge set. What is not so obvious is that it is minimal. See problems I36 and I37 to find out why.

Corollary *In any graph with n vertices and* $\delta \geq 1$, $\alpha' + \beta' = n$.

This requires just a little bit of explanation:

120 Show that the set constructed by the theorem contains $n - \alpha'$ edges. (Hint: First show that the number of edges is $\alpha' + (n - 2\alpha')$.)

121 How does problem 120, together with the theorem, prove the corollary?

More Problems

122 For each graph below, find α, α', and corresponding maximal independent sets.

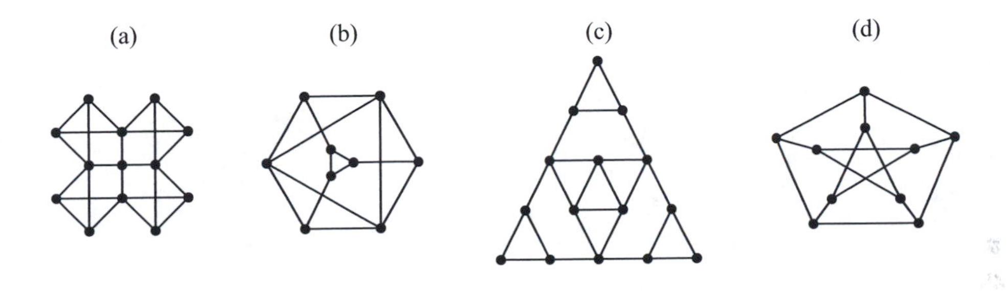

123 (a) Find three independent vertices in this graph.

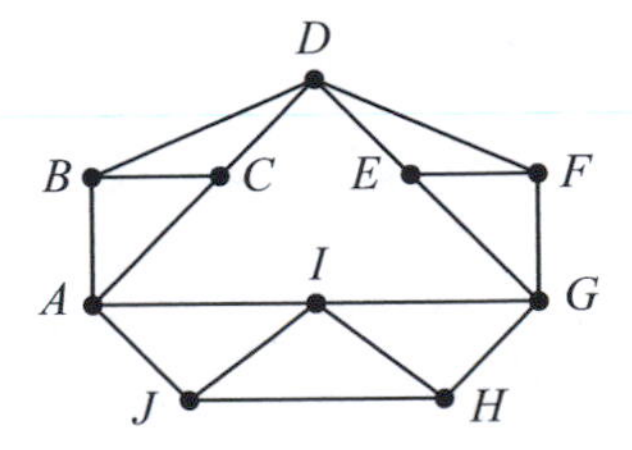

(b) Show that any independent vertex set in this graph can contain at most one of the vertices A, B, C, at most one of D, E, F, and at most one of G, H, I. Use this to conclude that if a set of four independent vertices exists, then it must include J.

(c) Show that any set of four independent vertices in this graph must include H.

(d) Prove that the independent set found in (a) is maximal.

124 (a) Find six independent vertices and four independent edges in this graph and explain how you can be certain that these sets are maximal without searching for larger independent sets.

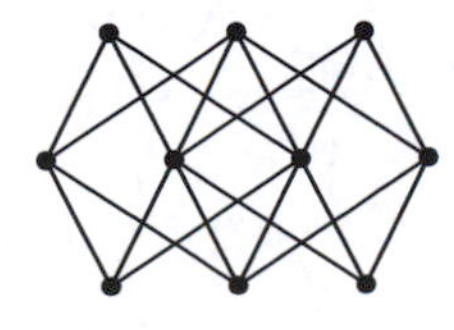

(b) Find eight independent vertices and six independent edges in this graph and explain how you can be certain that these sets are maximal.

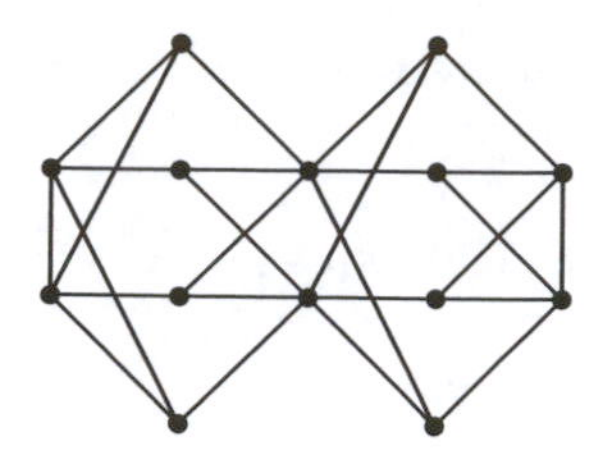

125 For each graph below, find α, α', and corresponding maximal independent sets.

(a)

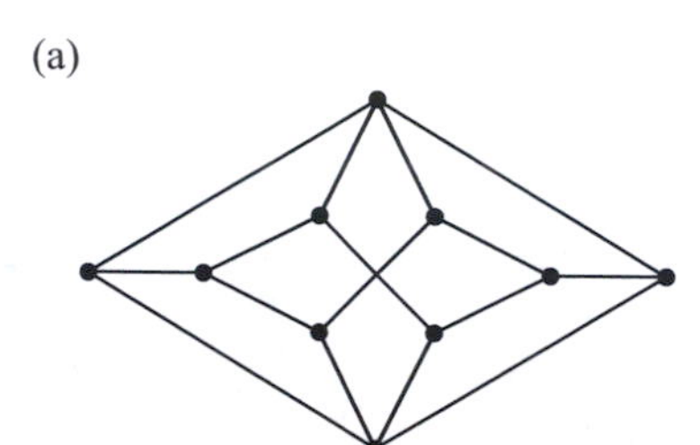

(b)

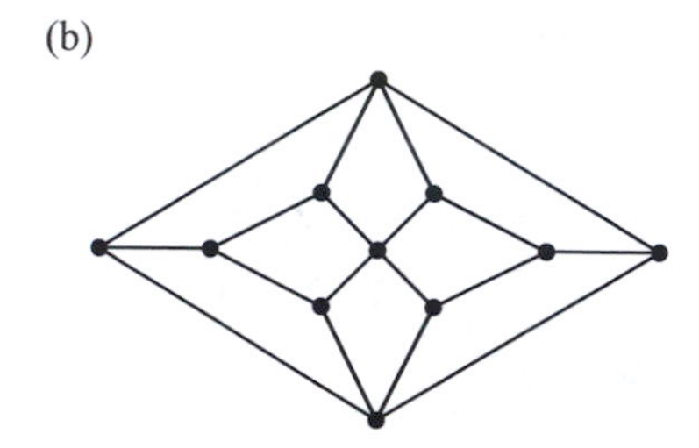

(c)

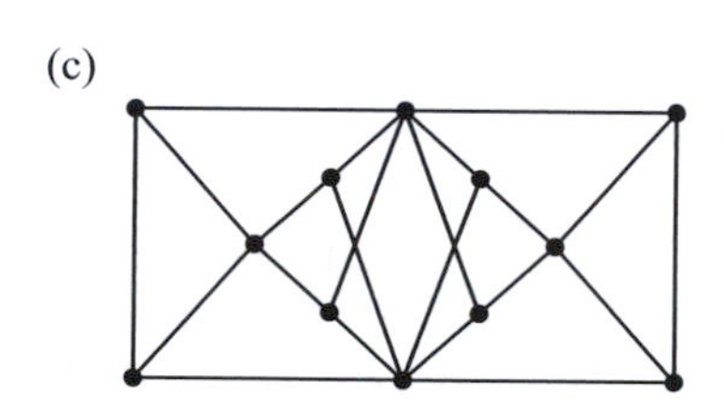

(d)

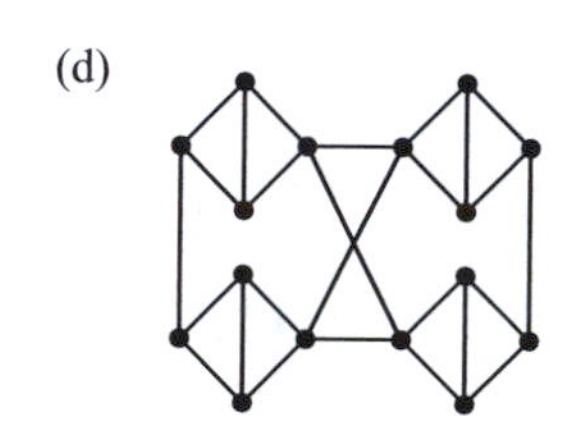

126 In a group of seven women and seven men, each woman likes exactly three of the men and each man likes exactly three of the women.

(a) Is there necessarily a way to pair up these people so that everyone is happy?

(b) Suppose that it goes both ways: whenever A likes B, then B also likes A. Then what is the answer? Why?

127 In any graph, a *clique* is a set of vertices that are all adjacent to each other. The *clique number* γ of a graph is the number of vertices in a maximal clique. Find γ for each graph below.

(a)

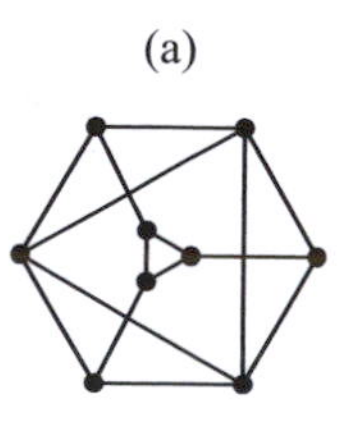

(b)

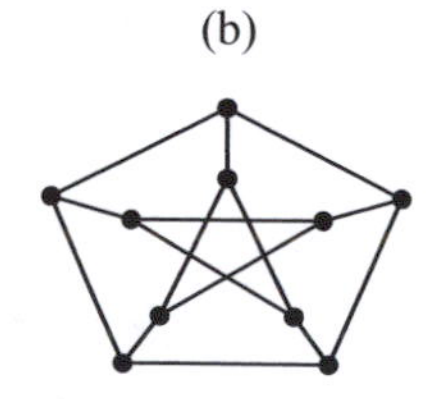

(c)

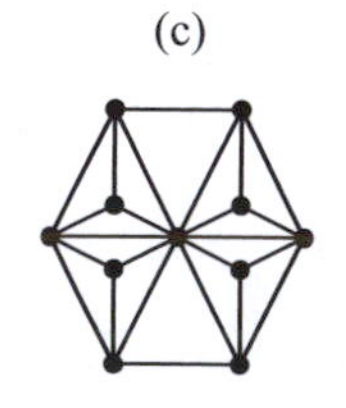

(d)

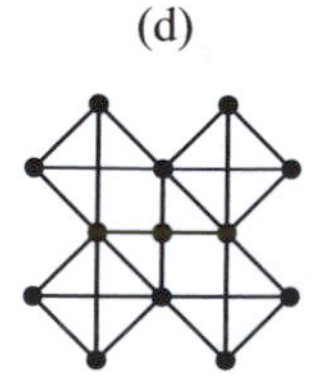

128 Show that if $\mathbf{G}_1$ and $\mathbf{G}_2$ are two complementary graphs, then α for $\mathbf{G}_1$ is equal to γ for $\mathbf{G}_2$.

129 For each graph below, which player has a winning strategy in the graph game described in this chapter? In each case, what is the strategy?

(a)

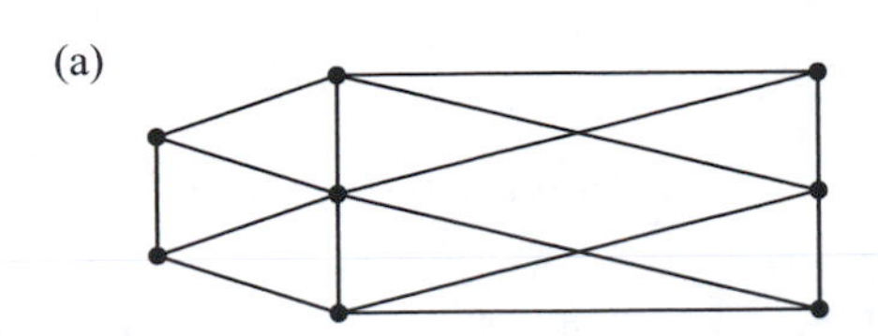

(b)

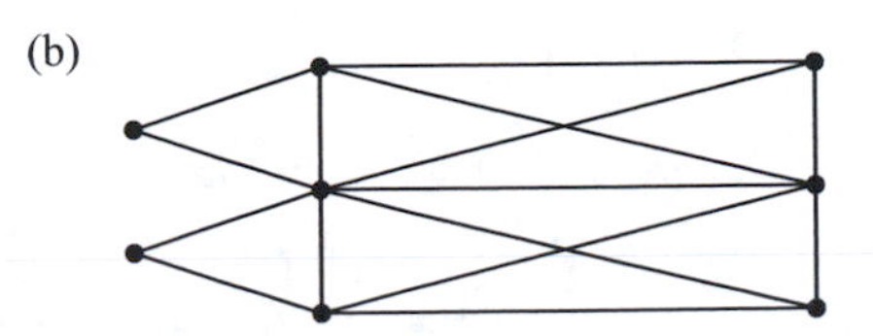

130 Use the covering theorems and answers to previous problems to find β, β' and corresponding minimal covering sets for each graph below.

(a)

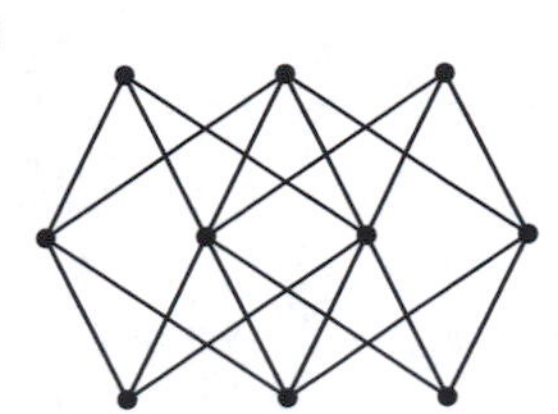

(b)

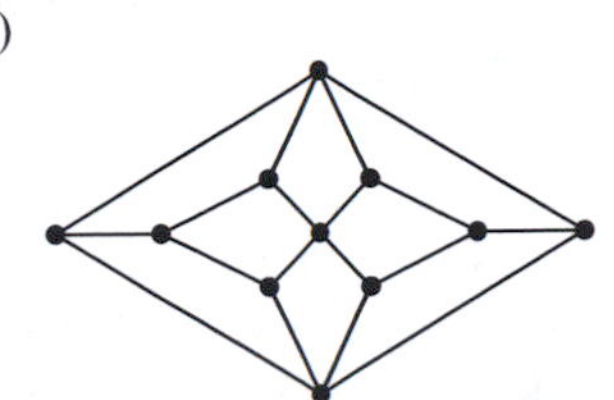

(c)

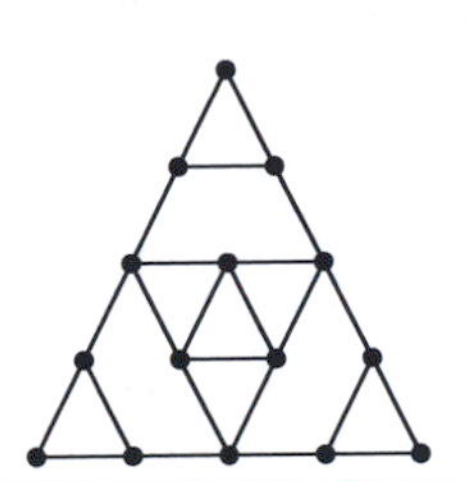

(d)

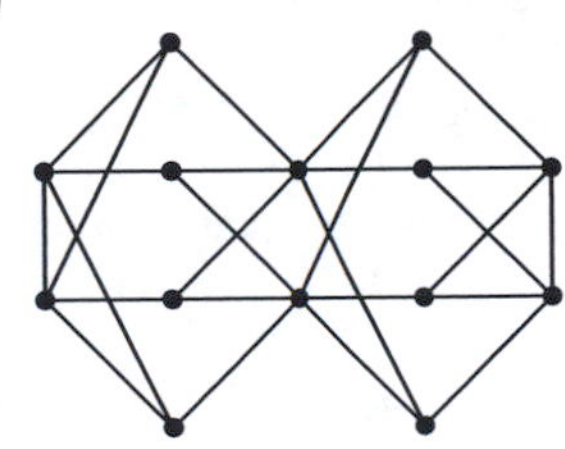

131 Use results from this section to derive each of the statements below for a graph with n vertices. Whenever β' is involved, assume that $\delta \geq 1$.

(a) $\alpha' \leq \beta$.

(b) $\alpha \leq \beta'$.

(c) $\beta + \beta' \geq n$.

(d) Equality holds in all of the above if the graph is bipartite.

(e) $\beta' \geq n/2$.

(f) $\beta \geq n/2$ if the graph is d-regular with $d \geq 1$.

(g) $\beta = \beta' = n/2$ if the graph is bipartite and d-regular with $d \geq 1$.

132 Which of the relations below are always true whenever **H** is a subgraph of **G**? For (d), assume $\delta \geq 1$ in both graphs.

(a) $\alpha(\mathbf{H}) \leq \alpha(\mathbf{G})$.

(b) $\alpha'(\mathbf{H}) \leq \alpha'(\mathbf{G})$.

(c) $\beta(\mathbf{H}) \leq \beta(\mathbf{G})$.

(d) $\beta'(\mathbf{H}) \leq \beta'(\mathbf{G})$.

133 Show that in any graph with e edges, $\alpha \leq e/\delta$ and $\beta \geq e/\Delta$. (Hint: Consider degree sums.)

Proof of Covering Theorem #1 Let **S** be a maximal independent vertex set in a given graph **G** and let **T** be the complement of **S**.

134 **T** must be a covering vertex set in **G**. Why?

135 To show that **T** is minimal, let **X** be any other covering vertex set in **G** and let k be the number of vertices in **X**.

(a) Prove that the complement of **X** is an independent vertex set. How many vertices are in this complement?

(b) Use (a) to show that $n - k \leq \alpha$.

(c) Use (b) to show that **X** contains at least $n - \alpha$ vertices. Explain why this proves that **T** is minimal.

Proof of Covering Theorem #2 We already know that the set of edges constructed by the theorem is a covering edge set. Call this set $\mathbf{T}'$. According to problem 120, $\mathbf{T}'$ contains exactly $n - \alpha'$ edges. To complete the proof we have to show that $\mathbf{T}'$ is minimal.

Let $\mathbf{X}'$ be any minimal covering edge set, and let **H** be the subgraph of **G** consisting of all n vertices of **G** and only the edges that are in $\mathbf{X}'$.

136 Show that $\mathbf{X}'$ is the only covering edge set in **H**.

In view of this, we can be sure that when Covering Theorem #2 is applied to **H**, the resulting covering edge set is $\mathbf{X}'$.

137 Complete the proof that $\mathbf{T}'$ is minimal by the following steps:

(a) Apply problem 120 to prove that the number of edges in $\mathbf{X}'$ is equal to $n - \alpha'(\mathbf{H})$, where $\alpha'(\mathbf{H})$ is the edge independence number for **H**.

(b) Can **H** contain more independent edges than **G**?

(c) Use the results of (a) and (b) to conclude that $\mathbf{T}'$ is minimal.

J

Connections and Obstructions

Any two vertices in a connected graph are joined by one or more paths. Now we will take into consideration the number of different paths that join two given vertices.

Internally Disjoint Paths

Let A and B be two vertices in a graph. We consider simple paths from A to B. Two or more paths of this type are called *internally disjoint* if they have nothing in common (vertices or edges) other than their endpoints.

J1 Find two internally disjoint paths from A to B in this graph and explain how you know that there do not exist three such paths.

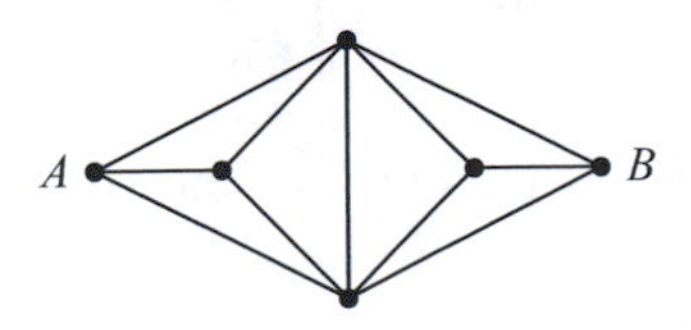

J2 Find as many internally disjoint paths as possible from A to B in each graph below.

(a)

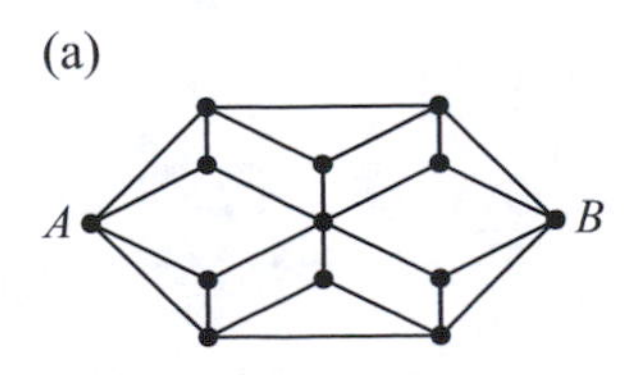

(b)

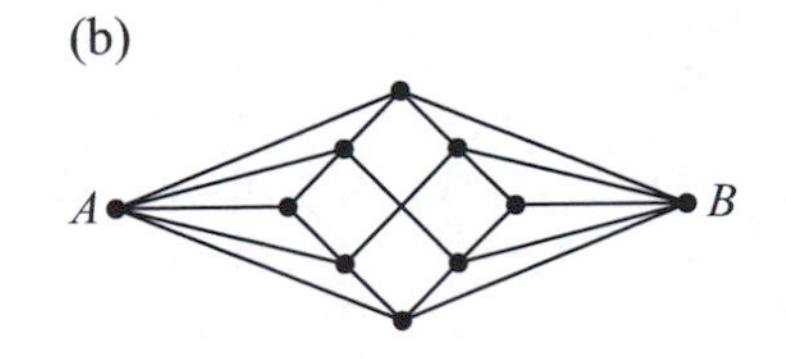

(c)

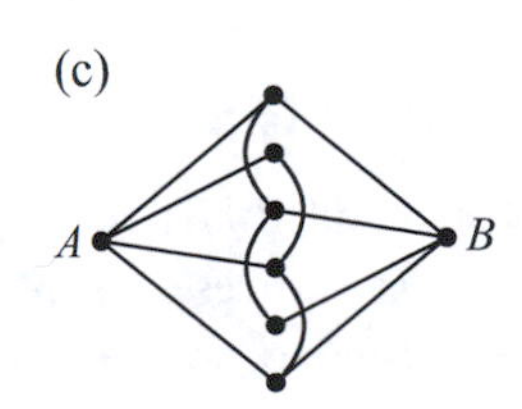

Edge-Disjoint Paths

A slightly different idea is to allow paths to have vertices in common, but not edges. Paths with no edges in common are called *edge-disjoint*.

J3 Find as many edge-disjoint paths as possible from A to B in each graph in problem J2.

J4 Decide whether each statement below is true or false.

(a) Edge-disjoint paths are always internally disjoint.

(b) Internally disjoint paths are always edge-disjoint.

Path Connection Numbers

For two vertices A and B in a graph, we define the *path connection numbers* $p(A, B)$ and $p'(A, B)$ in the following way:

$p(A, B)$ is the maximum number of internally disjoint paths from A to B, if any exist.
$p'(A, B)$ is the maximum number of edge-disjoint paths from A to B, if any exist.

J5 Explain why $p(A, B)$ can never be greater than $p'(A, B)$.

J6 Find $p(A, B)$ and $p'(A, B)$ in this graph.

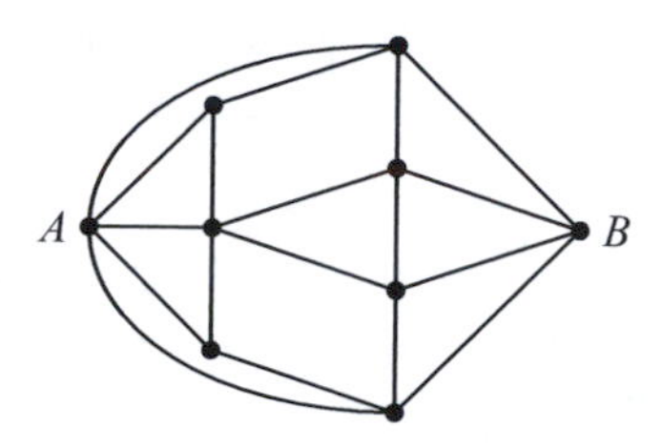

In cases where internally disjoint paths or edge-disjoint paths do not exist, there are two possibilities: If no path exists from A to B (that is, the vertices are in different components of the graph), then both $p(A, B)$ and $p'(A, B)$ are defined to be 0. If some path exists, but not two internally disjoint paths, then $p(A, B)$ is defined to be 1. If some path exists, but not two edge-disjoint paths, then $p'(A, B)$ is defined to be 1.

J7 If A and B are any two vertices in the complete graph $\mathbf{K}_5$, what are $p(A, B)$ and $p'(A, B)$? More generally, find $p(A, B)$ and $p'(A, B)$ if A and B are in $\mathbf{K}_n$.

J8 Explain how you know that $p(A, B)$ and $p'(A, B)$ are each less than or equal to $d(A)$, the degree of A, and similarly each is less than or equal to $d(B)$.

Blocking Sets

Let A and B be two vertices in a graph. A *blocking vertex set* (relative to A and B) is any set $\mathbf{S}$ of vertices such that every path from A to B includes at least one vertex in $\mathbf{S}$. (Assume $\mathbf{S}$ does not contain A or B.)

In other words, you can't go from A to B without passing through $\mathbf{S}$.

J9 Find a blocking vertex set relative to A and B consisting of three vertices in this graph and use this set to explain why there cannot exist more than three internally disjoint paths from A to B.

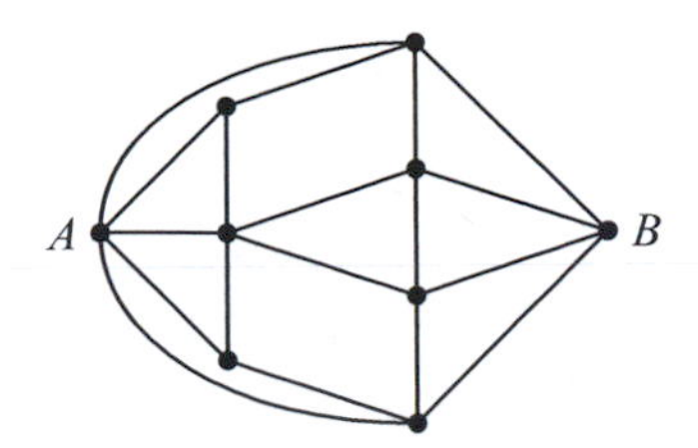

J10 Generalize the preceding problem to show that if a graph contains a blocking vertex set relative to A and B consisting of k vertices, then $p(A, B) \leq k$. Explain why this is true and use it to verify that you found the maximum number of internally disjoint paths from A to B in each graph in problem **J2**.

A *blocking edge set* (again, relative to A and B) is any set $\mathbf{S}'$ of edges such that every path from A to B includes at least one edge in $\mathbf{S}'$.

J11 There is a blocking edge set relative to A and B consisting of only two edges in this graph. Can you find it? Use this set to prove that $p'(A, B) = 2$.

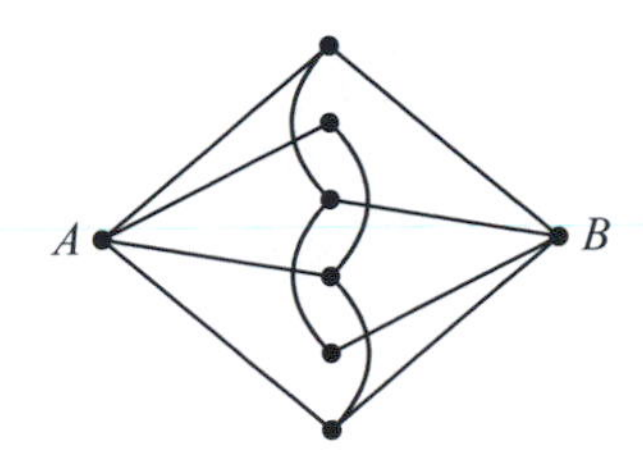

J12 State a version of the result in problem **J10** for edge-disjoint paths and blocking edge sets.

J13 Suppose that A and B are two vertices in the same component of a graph. Explain how you know that a blocking edge set, relative to A and B, always exists. Under what conditions would no blocking vertex set exist?

Problems **J10** and **J12** establish a relation between the path connection numbers $p(A, B)$ and $p'(A, B)$ and the number of elements in blocking vertex and edge sets. However the connection is even stronger than those problems indicate.

Menger's Theorem *Let A and B be two vertices in the same component of a graph. Then*

(1) If A and B are nonadjacent, then $p(A, B)$ is equal to the number of vertices in a smallest blocking vertex set relative to A and B.

(2) $p'(A, B)$ is equal to the number of edges in a smallest blocking edge set relative to A and B.

J14 (a) Find four internally disjoint paths from A to B in this graph.

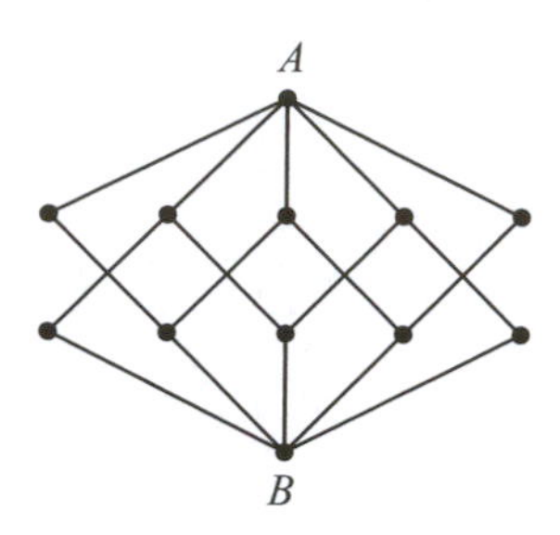

(b) According to part (1) of Menger's Theorem, there must exist either five internally disjoint paths from A to B in the graph above or else there is a blocking vertex set relative to A and B consisting of only four vertices. Which is it? Find either the five paths or the four vertices.

J15 (a) Find four edge-disjoint paths from A to B in this graph.

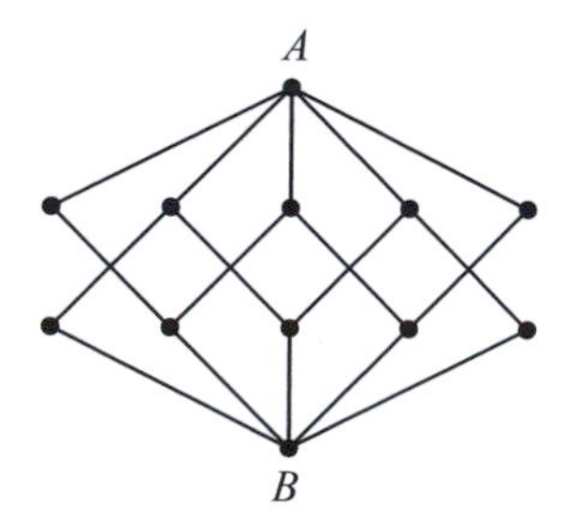

(b) According to part (2) of Menger's Theorem, there must exist either five edge-disjoint paths from A to B in the graph above or else there is a blocking edge set relative to A and B consisting of only four vertices. Which is it? Find either the five paths or the four edges.

The proof of Menger's Theorem depends on network flow theory and will not be included here.

k-Connected Graphs

Some connected graphs are, in a certain sense, more connected than others. This idea is based on path connection numbers:

Let **G** be a graph with two or more vertices, and let k be a positive integer. **G** is called *k-connected* if there exist at least k internally disjoint paths from A to B for each pair of vertices A and B in **G**. In symbols, this condition can be written as $p(A, B) \geq k$.

Notice that a 1-connected graph is exactly the same thing as a connected graph: $p(A, B) \geq 1$ for all vertices A and B.

J16 Explain why any graph that contains a Hamilton cycle must be 2-connected.

J17 Show that this graph is 3-connected by considering different combinations of vertices A and B.

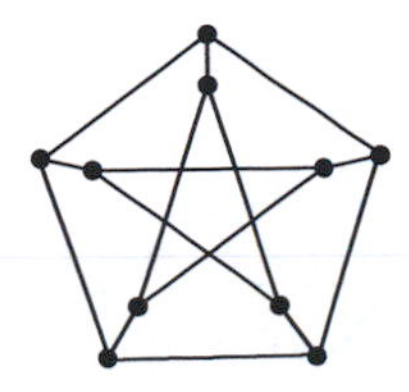

J18 Find a 3-regular graph that is 2-connected but not 3-connected. (Suggestion: Start with a cycle of length 8 and add four edges.)

Vertex Cut Sets and Vertex Cut Numbers

In any connected graph, a *vertex cut set* is a set of vertices which, when removed, result in a disconnected graph: That is, after the members of the vertex cut set are removed, the remaining graph is disconnected.

Remember that when vertices are removed from a graph, any edges that have these vertices as endpoints are also removed.

J19 Find a smallest possible vertex cut set in the graph of problem **J17**. (Hint: Any vertex cut set is a blocking vertex set relative to some pair of vertices.)

J20 Explain why a complete graph has no vertex cut sets, but any incomplete connected graph contains at least one vertex cut set. (Hint: Remove all but two vertices.)

The vertex cut number of a graph

For any incomplete connected graph **G**, we define the *vertex cut number* $q(\mathbf{G})$ to be the minimum number of vertices in any vertex cut set in **G**. For example, we saw that $q(\mathbf{G}) = 3$ for the graph in problem **J17**.

J21 Let **G** be an incomplete connected graph in which A is a vertex of minimum degree δ.

(a) Show that there must exist a vertex B in **G** that is nonadjacent to A.

(b) Show that **G** contains a vertex cut set consisting of δ vertices.

J22 Use the result of the preceding problem to show that the vertex cut number of an incomplete connected graph always satisfies the inequality $q(\mathbf{G}) \leq \delta$. Find an example in which $q(\mathbf{G}) < \delta$.

The vertex cut number of a graph can be used to determine whether the graph is k-connected, for any given k:

Corollary to Menger's Theorem *Let* **G** *be an incomplete connected graph. Then* **G** *is k-connected if and only if $q(\mathbf{G}) \geq k$.*

J23 Show how this corollary follows from Menger's Theorem by the following steps.

(a) Explain why, for any two vertices A and B in G, there cannot exist a blocking vertex set relative to A and B containing fewer than $q(\mathbf{G})$ vertices.

(b) For any two vertices A and B, $p(A, B) \geq q(\mathbf{G})$. (Use Menger's Theorem here.)

(c) If $q(\mathbf{G}) \geq k$, then $\mathbf{G}$ is k-connected.
(d) There exist vertices A and B in $\mathbf{G}$ and a blocking vertex set relative to A and B containing exactly $q(\mathbf{G})$ vertices.
(e) If $q(\mathbf{G}) < k$, then $\mathbf{G}$ is not k-connected.

More Problems

J24 (a) Find three internally disjoint paths from A to B in this graph.

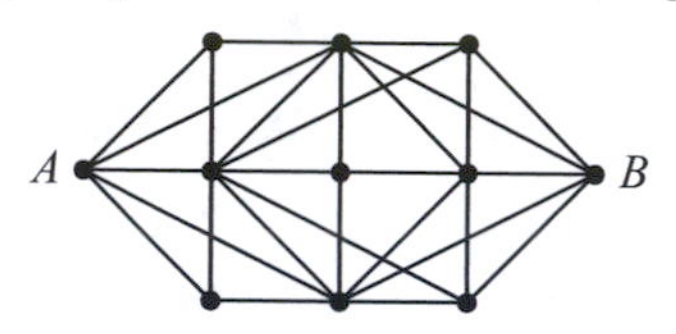

(b) Find either four internally disjoint paths from A to B in the graph above or a blocking vertex set relative to A and B consisting of three vertices.

J25 Find five edge-disjoint paths from A to B in the graph in problem J24.

J26 (a) Find three internally disjoint paths from A to B in this graph.

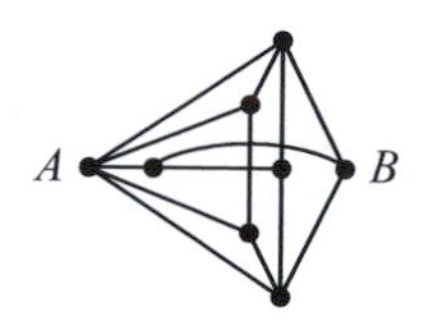

(b) Find either four internally disjoint paths from A to B in the graph above or a blocking vertex set relative to A and B consisting of three vertices.

J27 Find a maximal set of edge-disjoint paths from A to B in the graph in problem J26, and a minimal blocking edge set relative to A and B.

J28 (a) Find four internally disjoint paths from A to B in this graph.

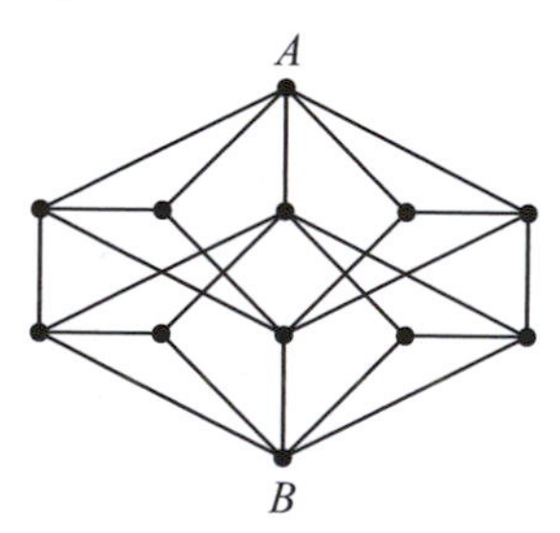

(b) Find either five internally disjoint paths from A to B in the graph above or a blocking vertex set relative to A and B consisting of four vertices.

J29 Find a maximal set of edge-disjoint paths from A to B in the graph in problem J28, and a minimal blocking edge set relative to A and B.

J30 In this graph, find either five edge-disjoint paths from A to B or a blocking edge set relative to A and B consisting of four edges.

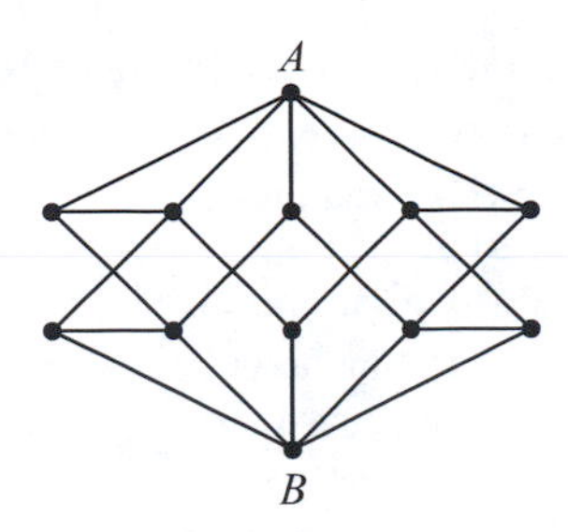

J31 Suppose that in a certain graph, A is a vertex of degree 5, B is a vertex of degree 4, and $p(A, B) = 3$. Then what are the possible values of $p'(A, B)$?

J32 Make up a reasonable definition for an *edge cut set* in a connected graph. Does every connected graph contain an edge cut set? (Watch out for small graphs.)

J33 Define the *edge cut number* $q'(\mathbf{G})$ for a given connected graph in the obvious way, and show that $q'(\mathbf{G}) \leq \delta$.

J34 In a connected graph with two or more vertices, suppose that every edge is contained in some cycle. Show that the edge cut number $q'(\mathbf{G})$ for this graph must be at least 2. Is a similar result true for the vertex cut number if every vertex is contained in a cycle?

J35 Find an example of a graph in which $q(\mathbf{G}) = 1$, $q'(\mathbf{G}) = 2$, and $\delta = 3$.

J36 Find an example of a graph in which $q(\mathbf{G}) = 2$, $q'(\mathbf{G}) = 3$, and $\delta = 4$.

In the following problems we will show that the cut numbers in any incomplete connected graph $\mathbf{G}$ satisfy the relation $q(\mathbf{G}) \leq q'(\mathbf{G})$. Combining this with the result of problem J33, we have $q(\mathbf{G}) \leq q'(\mathbf{G}) \leq \delta$.

Let $\mathbf{G}$ be an incomplete connected graph with n vertices, and let $\mathbf{S}'$ be a minimal edge cut set. $\mathbf{S}'$ contains $q'(\mathbf{G})$ edges. The idea of the proof is to construct a vertex set $\mathbf{S}$ by selecting one endpoint from each edge of $\mathbf{S}'$. Then $\mathbf{S}$ will contain at most $q'(\mathbf{G})$ vertices. If $\mathbf{S}$ is a vertex cut set, then it will follow that $q(\mathbf{G}) \leq q'(\mathbf{G})$.

J37 Explain how that will follow, assuming that $\mathbf{S}$ is a vertex cut set.

J38 Something is wrong with the following argument. What is it?

To show that $\mathbf{S}$ is a vertex cut set, we must show that $\mathbf{G}$ becomes disconnected when the vertices of $\mathbf{S}$ are removed. When the vertices in $\mathbf{S}$ are removed, so are all edges that go to those vertices. These include all of the edges in $\mathbf{S}'$. Since $\mathbf{S}'$ is an edge cut set, $\mathbf{G}$ becomes disconnected when the edges of $\mathbf{S}'$ are removed. Therefore $\mathbf{G}$ becomes disconnected when the vertices of $\mathbf{S}$ are removed. (For example, see what goes wrong when this argument is applied to the graph below with $\mathbf{S}$ consisting of the three vertices at the left and $\mathbf{S}'$ consisting of the three horizontal edges.)

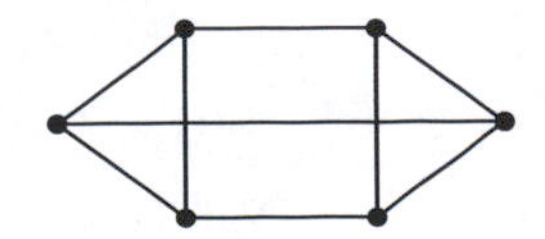

To get around this problem, we have to select the vertices in **S** a little more carefully. In particular, we single out two vertices that will never be selected and that end up in different components after the vertices in **S** are removed.

J39 Show that when the edges of **S**′ are removed from **G**, the vertices can be separated into two subsets **U** and **V** such that no edges go from **U** to **V**.

J40 Let n_1 and n_2 be the numbers of vertices in **U** and **V**, respectively. Then $n_1 + n_2 = n$. Prove that $n_1 n_2 \geq n - 1$. (Suggestion: Consider the product $(n_1 - 1)(n_2 - 1)$. What must be true about this number?)

J41 Show that the minimum degree in **G** is at most $n - 2$. Conclude that $q'(\mathbf{G}) \leq n - 2$.

J42 Show that there must exist at least one vertex u in **U** and one vertex v in **V** such that u and v are not endpoints of the same edge in **S**′. Use this fact to complete the proof that $q(\mathbf{G}) \leq q'(\mathbf{G})$.

J43 Give another proof that $q(\mathbf{G}) \leq q'(\mathbf{G})$ based on Menger's Theorem. Suggestion: First show that $q'(\mathbf{G})$ is the number of edges in a minimal blocking edge set relative to some A and B. Then use Menger's Theorem to show that there is a blocking vertex set containing at most $q'(\mathbf{G})$ vertices.

J44 Prove that in any regular, connected, bipartite graph with three or more vertices, the vertex cut number $q(\mathbf{G})$ must be at least 2. (Hint: Show that if $q(\mathbf{G}) = 1$, then **G** contains a subgraph that contradicts the result of problem C24.)

J45 Let **G** be a connected graph with two or more vertices, and suppose that every vertex in **G** has even degree.

(a) Show that the edge cut number $q'(\mathbf{G})$ cannot be 1.

(b) Show that $q'(\mathbf{G})$ must be even.

J46 A graph **G** is called k-edge-connected if there exist at least k edge-disjoint paths from A to B for each pair of vertices A and B in **G**. Equivalently, $p'(A, B) \geq k$. State and prove a modified version of the corollary to Menger's Theorem that applies to k-edge-connected graphs.

Vertex Coloring

In this section we consider the problem of coloring the vertices of a graph.

Definition A *proper vertex coloring* of a graph is the assignment of a color (usually represented by a number or some other symbol) to each vertex of the graph in such a way that no color occurs at adjacent vertices. Another way of stating this condition is that *each edge has two endpoints of different colors*. Still another way to say it is that *the set of vertices of any one color is an independent vertex set.*

Example

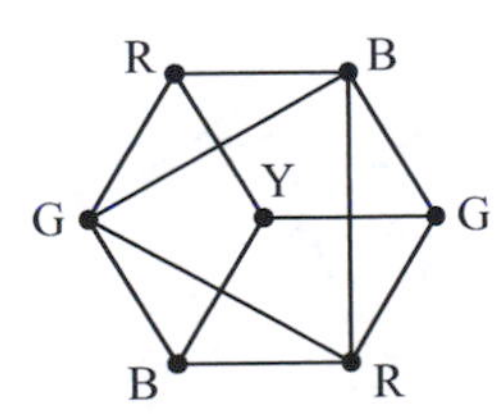

R = red
B = blue
G = green
Y = yellow

The Vertex Coloring Number of a Graph

We are interested in coloring the vertices of a given graph **G** properly using the smallest possible number of different colors. Such a coloring is called a *minimal proper vertex coloring* of **G**. The *vertex coloring number* χ of **G** is the number of colors that occur in a minimal proper vertex coloring.

K1 Show that the coloring in the example above is minimal. In other words, show that $\chi = 4$ for this graph. Do this by the following steps:

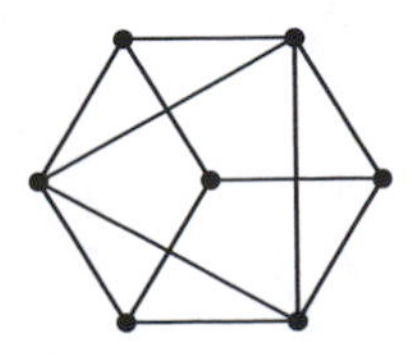

(a) Begin by coloring the vertices of a 3-cycle. (These colors must all be different, and it doesn't matter what they are called.)

(b) Continue coloring vertices properly, always trying to avoid introducing a new color and looking for new vertices for which there is only one possible choice of a color.

(c) Finally arrive at a vertex that requires a fourth color.

K2 What is the value of χ if **G** is a bipartite graph? Consider two cases separately:

(a) If **G** contains at least one edge;

(b) If **G** contains no edges.

K3 Find a minimal proper vertex coloring for each graph below.

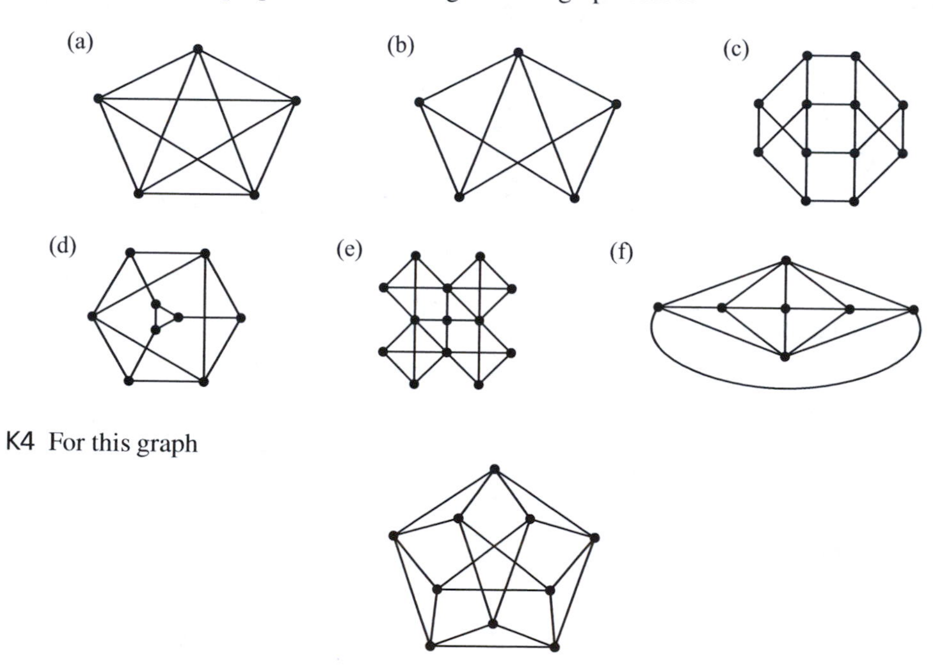

K4 For this graph

(a) Show that in any proper vertex coloring, no color can occur more than three times. (Hint: What is α for this graph?)

(b) Use (a) to show that three colors could not be enough for a proper vertex coloring of this graph.

(c) Find a proper vertex coloring using four colors.

Vertex Coloring Theorems

The preceding problem suggests a general relation that holds between the values of χ and α for any graph.

K5 Fill in the blanks appropriately: In any graph with vertex independence number α and vertex coloring number χ, the number of vertices of any particular color is at most _____, and therefore the total number of vertices in the graph is at most _____.

Vertex Coloring Theorem #1 *In any graph with n vertices and vertex independence number α, the vertex coloring number χ satisfies the inequality $\chi \geq n/\alpha$.*

K6 Show how this result follows from problem K5.

Vertex Coloring Theorem #1 places a lower limit on the value of χ for a given graph. Next we look at some theorems that establish upper limits for χ. These are related to particular coloring strategies. In general, we will construct a proper vertex coloring of a graph by coloring vertices one at a time, introducing a new color only when it is necessary to avoid creating adjacent vertices of the same color.

K7 Try that for each graph below, coloring the vertices in the indicated order. In which cases does this result in a minimal coloring?

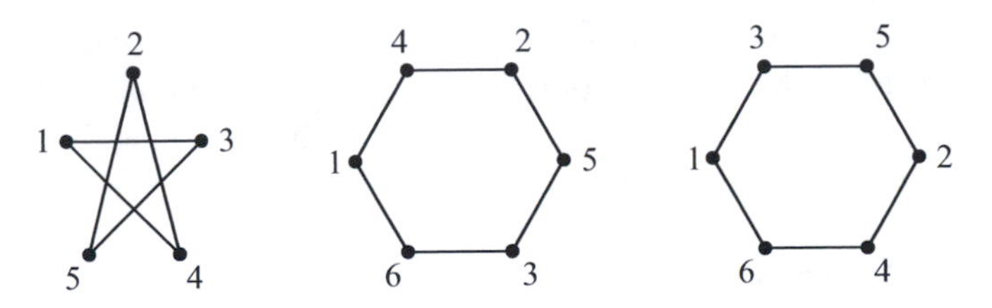

K8 Apply the same procedure to this graph, starting with the vertices labeled 1, 2, and 3.

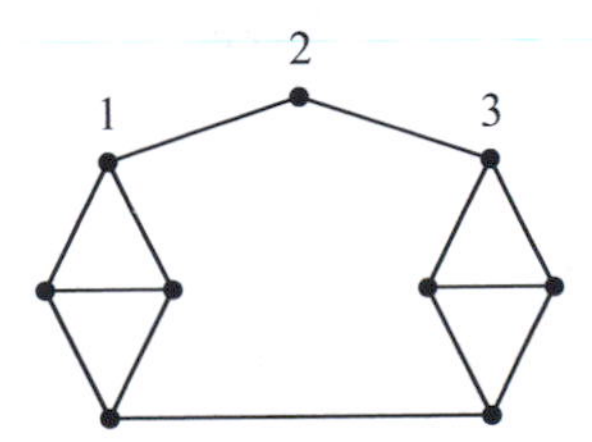

K9 Explain why χ is never greater than 4 for a graph in which the maximum degree is 3. What must be true for a graph in which the maximum degree is 100?

The same reasoning shows that $\chi \leq \Delta + 1$ for every graph, where as usual Δ is the maximum degree.

K10 Of the following graphs, which ones have $\chi = \Delta + 1$? $\mathbf{K}_5$, $\mathbf{K}_6$, $\mathbf{C}_5$, $\mathbf{C}_6$.

The relation between χ and Δ is summarized in the theorem below. The second part, which is known as Brooks' Theorem, will be proved in the problems below and at the end of the chapter.

Vertex Coloring Theorem #2

(1) *For every graph, $\chi \leq \Delta + 1$.*

(2) *For every connected graph that is not complete and not an odd cycle graph, $\chi \leq \Delta$.*

By an odd cycle graph, we mean $\mathbf{C}_n$ where n is odd.

K11 See what this theorem tells you about the graph in problem K4.

K12 If a graph is disconnected and you know the value of χ for each component, how would you determine χ for the entire graph?

K13 Returning to the graph in problem K8, see what happens when you color the vertices in this order:

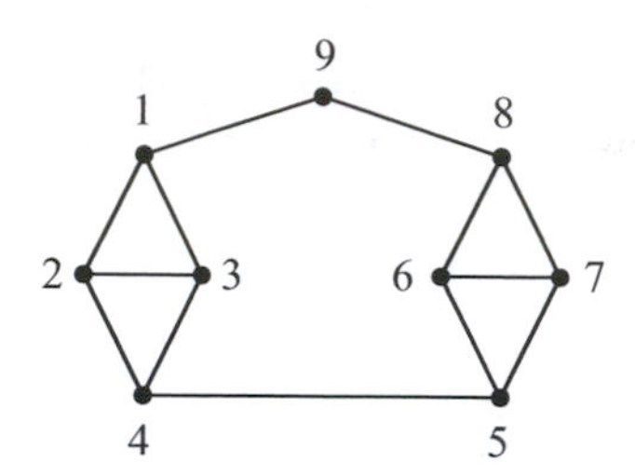

Problem K13 suggests how Brooks' Theorem might be proved in the case where the graph is not regular. The idea is to order the vertices so that every time a new vertex is reached, fewer than Δ adjacent vertices have already been colored. If this can be done, then it will never be necessary to use more than Δ colors.

K14 Find an ordering of the vertices in this graph so that

(1) Every vertex other than the last is adjacent to some vertex that comes later in the ordering, and

(2) The final vertex has non-maximal degree.

If a graph is connected and non-regular, it is always possible to order the vertices so that conditions (1) and (2) are satisfied. An easy way to do this is to first order the vertices starting at a vertex of non-maximal degree and follow paths to new vertices, always selecting a vertex adjacent to one that has already been reached. Since the graph is connected, all vertices are eventually reached in this way. Finally, reverse this order and the result will satisfy (1) and (2). Try this for the graph in problem K14.

K15 Explain why no more than Δ colors are used when vertices are colored in an order satisfying conditions (1) and (2).

Since every non-regular connected graph has such an ordering, this proves Brooks' Theorem for graphs of that type. The proof for regular graphs will be given in problems K53–K61.

In some graphs, better results can be obtained from different orderings of the vertices. In particular, we look at what happens when vertices are colored in order of their degrees, from highest to lowest.

K16 Try that on this graph, introducing new colors only when necessary. Does this result in a minimal coloring?

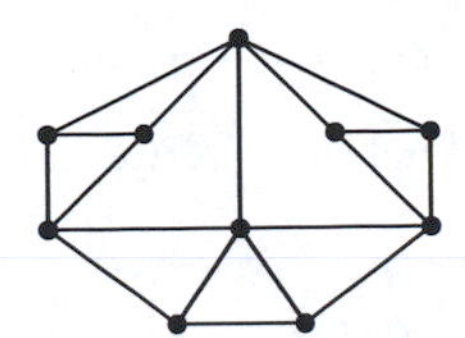

K17 Repeat problem K16, coloring from highest to lowest degree, except this time assign different colors to each of the first four vertices even if this is not necessary to avoid adjacent vertices of the same color. Why are no new colors needed after the first four vertices?

The two preceding problems suggest how the degree sequence of a graph can lead to information about the coloring number. The conclusion $\chi \leq 4$ can be reached for any graph with degree sequence $(5, 5, 4, 4, 3, 3, 3, 3, 3, 3)$. One way to explain this is by the coloring rule given in problem K17: After four colors are assigned to the first four vertices, no new colors are ever needed. But the coloring rule in problem K16 also shows that $\chi \leq 4$: In that case, new colors may be needed after the first four vertices have been colored, but there is never a need for a fifth color.

Using the same reasoning as above, we can prove this:

Vertex Coloring Theorem #3 *Suppose that for a graph* **G**, *there is a number k such that* **G** *contains at most k vertices of degree k or greater. Then* **G** *has a proper vertex coloring using at most k colors.*

In practice, it is easier to apply this theorem in the form of an algorithm that produces the appropriate number k:

Algorithm form of Vertex Coloring Theorem #3: The Upper Bound Algorithm for χ

Arrange the degrees of a graph in decreasing order: $d_1 \geq d_2 \geq d_3 \cdots$. Then place the integers $0, 1, 2, \ldots$ directly under these degrees until you reach an integer k that is strictly greater than the degree above it:

d_1	d_2	d_3	$\cdots$	d_{k+1}	$\cdots$
0	1	2		k	**Stop** when $k > d_{k+1}$

Then for this graph, $\chi \leq k$.

Example For the degree sequence of the graph in problem K16,

5	5	4	4	3	3	3	3	3	3
0	1	2	3	4	**Stop**				

Conclusion: $\chi \leq 4$.

K18 In the case of a regular graph, does the Upper Bound Algorithm tell us anything we didn't already know?

Why the algorithm works

The Upper Bound Algorithm follows from Vertex Coloring Theorem #3. This is shown in the next problem.

K19 Let k be the number found by the Upper Bound Algorithm. Show that **G** contains at most k vertices of degree k or greater. (Remember that the degrees are in decreasing order.)

The Four Color Theorem

Finally, we state the most famous vertex coloring theorem of all, and the most difficult to prove:

Vertex Coloring Theorem #4: The Four Color Theorem *For every planar graph,* $\chi \leq 4$.

After more than 120 years of unsuccessful attempts, this was finally proved in 1976 by Kenneth Appel and Wolfgang Haken at the University of Illinois by a method that required extensive checking of details by computer. Approximately 1200 hours of computer time was needed.

It is surprisingly easy to establish a weaker version of this theorem, that $\chi \leq 5$ for every planar graph (the Five Color Theorem), and even easier to show that $\chi \leq 6$ (the Six Color Theorem). Starting with the latter, we will prove both of these here.

Proof of the Six Color Theorem

The proof is by mathematical induction on the number of vertices in the graph. Alternatively, it can be stated as a proof by contradiction, as explained below. In either case, we begin with a planar graph **G** and assume that every smaller planar graph (one with fewer vertices) has a proper vertex coloring that uses no more than six colors. The key to the proof is that **G** must contain at least one vertex of degree ≤ 5, as was proved for every planar graph in problem H31. Let U be any such vertex and consider the subgraph **H** of **G** formed by removing U, and all edges at U, from **G**.

K20 We know that **H** has a proper vertex coloring using at most six colors. Why?

K21 It is now possible to color vertex U to produce the required coloring of the graph **G**. How do we know this?

Problems K20 and K21 show that **G** has a proper vertex coloring that uses no more than six colors if every smaller planar graph has such a coloring. It follows by mathematical induction that the same is true for all planar graphs. Or to put it another way, suppose that some planar graph exists for which $\chi > 6$. Then a smallest such graph (one with the fewest vertices) exists.

K22 Use problems H31 and K21 to obtain a contradiction.

Proof of the Five Color Theorem

K23 Try to use the same argument as above, starting with a planar graph **G** and assuming that every smaller planar graph has a proper vertex coloring that uses no more than five colors. What difficulty can occur when you try to color vertex U? Describe the situation in which there is a problem.

In order to overcome the difficulty encountered in problem K23, we introduce the concept of a *color switch*. This is a process in which two colors are interchanged on a set of vertices without creating any adjacent vertices of the same color.

K24 Find a red-blue color switch on a set of three vertices in the graph below. Is there more than one possibility?

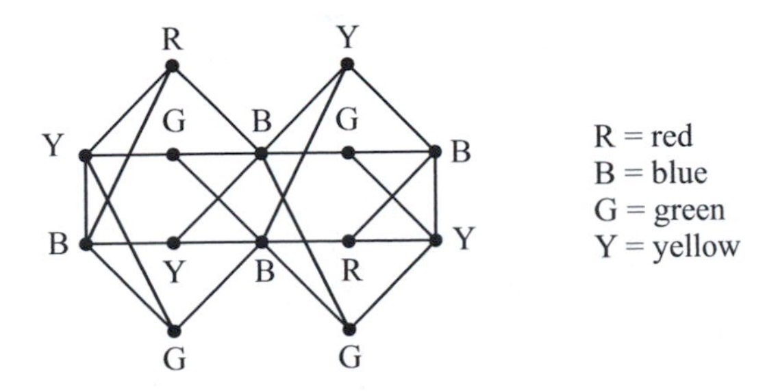

K25 Show that only one blue-green color switch is possible in the graph above. Explain why this is true using the term "connected".

Color switch

For any proper vertex coloring of a graph **G**, fix two colors and let **H** be the subgraph of **G** consisting of all vertices at which these colors occur and all edges joining vertices of these colors. Then a color switch is possible on any connected component of **H**.

Recall the difficulty encountered in problem K23. An example of this occurs in the graph below, in which five colors occur at vertices adjacent to the uncolored vertex U.

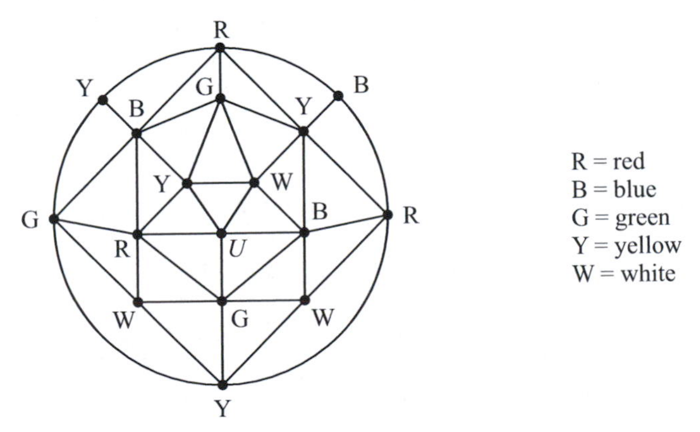

In order to color U without introducing a sixth color, we look for a color switch that reduces the number of colors that occur at vertices adjacent to U.

K26 Try a red-blue switch. What goes wrong?

K27 (a) Find a green-yellow switch that changes the green vertex adjacent to U but not the yellow one. What does this allow you to do?

(b) Find a green-yellow switch that changes the yellow vertex adjacent to U but not the green one.

K28 Repeat problem K27 using green and white.

Returning now to the proof of the Five Color Theorem, we assume this arrangement of colors at vertices adjacent to U:

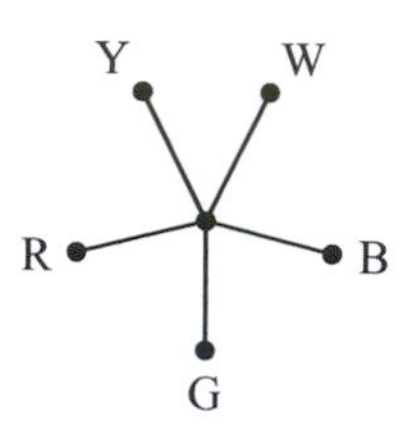

K29 Show that either a red-blue switch or a green-yellow switch reduces the number of colors that occur at vertices adjacent to U. (Hint: If a red-blue switch doesn't work, show that there must be a red-blue path that, together with U, forms a cycle separating the green neighbor of U from the yellow one. Remember that this occurs in a planar graph.)

K30 How does that prove the Five Color Theorem?

Map Coloring

When coloring regions on a map, two regions that have a common border have to be colored with different colors. However, this restriction does not apply to regions that meet at only a single point. Below are some examples,

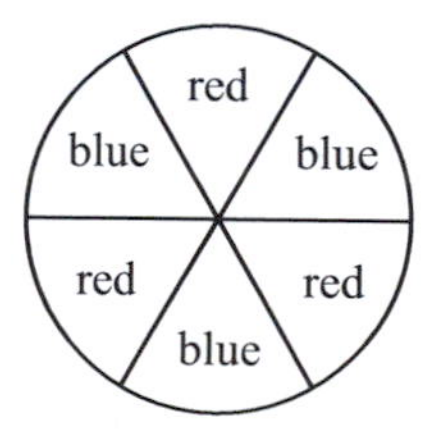

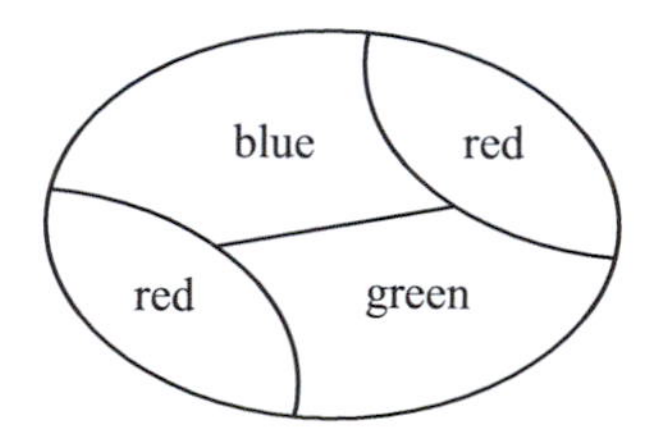

As a consequence of the Four Color Theorem, we have this:

Corollary (Map Coloring Theorem) Every map can be colored using at most four colors.

K31 How does this follow from Vertex Coloring Theorem #4? What does this have to do with coloring vertices on a graph? Here is a generous hint:

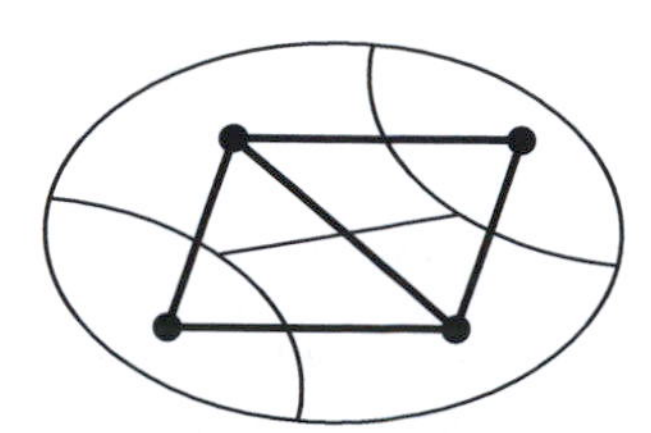

K32 Color this map by finding a minimal proper vertex coloring in an appropriate graph.

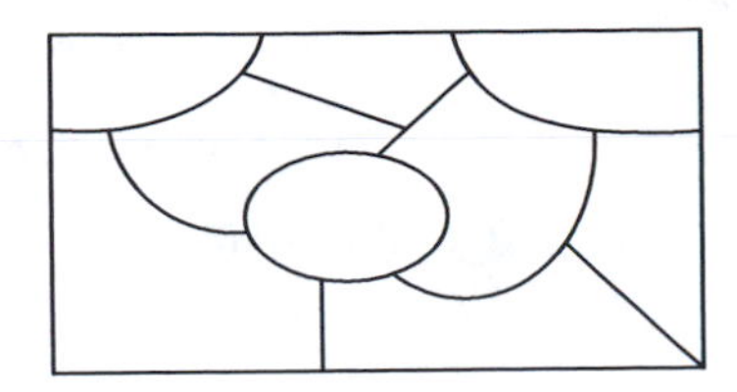

More Problems

K33 Find a minimal proper vertex coloring for each graph below.

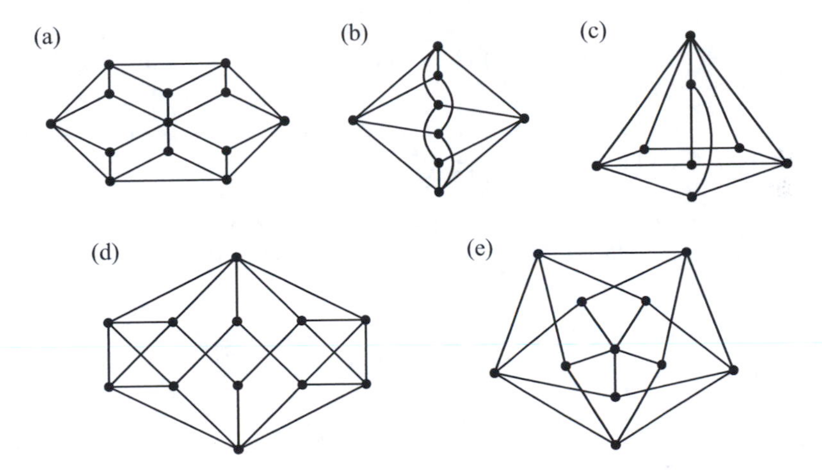

K34 For each graph below, one of the vertex coloring theorems says that $\chi \leq 4$. In each case indicate which theorem applies and find a minimal proper vertex coloring.

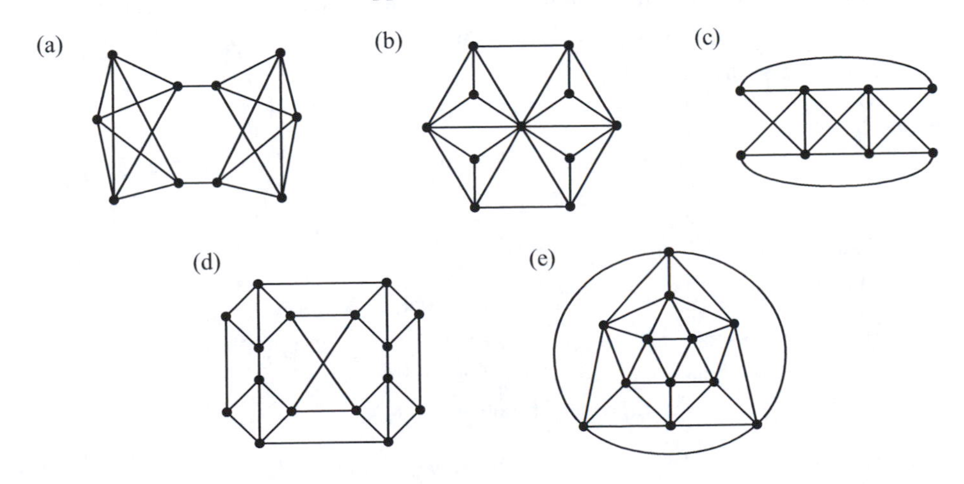

K35 Let $\mathbf{H}$ be a subgraph of $\mathbf{G}$. Is it necessarily true that $\chi(\mathbf{H}) \leq \chi(\mathbf{G})$? Explain.

K36 Suppose that a graph $\mathbf{G}$ is the union of two subgraphs $\mathbf{G}_1$ and $\mathbf{G}_2$ that have only a single vertex in common. Show that $\chi(\mathbf{G})$ is the maximum of $\chi(\mathbf{G}_1)$ and $\chi(\mathbf{G}_2)$.

K37 Apply the Upper Bound Algorithm to each degree sequence below. In each case, what is the conclusion?

(a) $(1,2,3,3,4,4,4,5)$ (c) $(7,7,7,7,7,5,5,5,5,5)$

(b) $(2,2,3,4,3,2,2)$ (d) $(8,7,6,5,4,3,3,3,3)$

K38 (a) Apply the Upper Bound Algorithm to this graph.

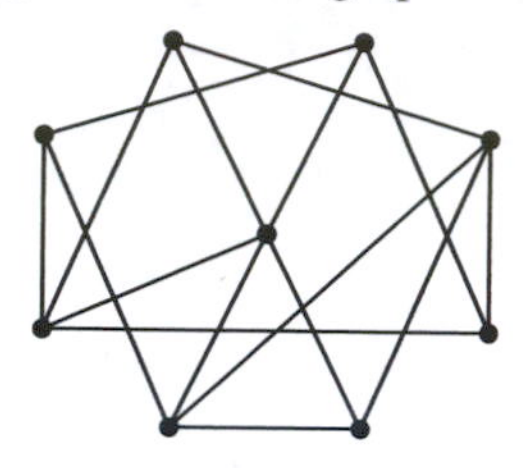

(b) Color the vertices from highest to lowest degree, assigning different colors to each of the first four vertices.

(c) Now try to achieve a better result by introducing new colors only when necessary. Does this mean the Upper Bound Algorithm is wrong?

K39 Below is a proper vertex coloring using the usual five colors.

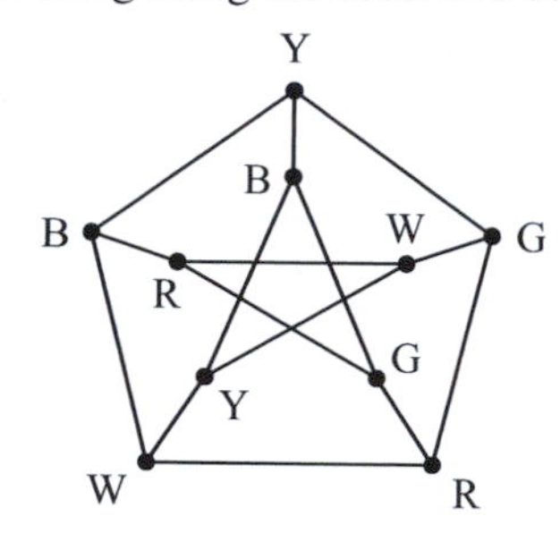

(a) Select any color and eliminate it by re-coloring all vertices of that color.

(b) Next try to eliminate a second color in the same way.

(c) What is the value of χ for this graph?

K40 Show that any graph with vertex coloring number χ must contain at least χ vertices of degree $\geq \chi - 1$. (Hint: If not, then some color occurs only at vertices of degree $\leq \chi - 2$. Show that this color can be eliminated.)

K41 Use the result of the preceding problem to show that for any graph, $\chi(\chi - 1) \leq 2e$, where e is the number of edges.

K42 (a) Show that in a minimal proper vertex coloring of a graph, any two colors must occur at adjacent vertices somewhere in the graph.

(b) Use (a) to give another proof that $\chi(\chi - 1) \leq 2e$.

K43 Show that if $\mathbf{G}_1$ and $\mathbf{G}_2$ are complementary graphs, then the sum $\chi(\mathbf{G}_1) + \chi(\mathbf{G}_2)$ is at most $n + 1$, where n is the number of vertices in each graph. (Suggestion: Let $r = \chi(\mathbf{G}_1)$, $s = \chi(\mathbf{G}_2)$, and suppose that $r + s \geq n + 2$. Use the result of problem K40 to show that there must be some vertex that has degree $\geq r - 1$ in $\mathbf{G}_1$ and degree $\geq s - 1$ in $\mathbf{G}_2$. Why is that a contradiction?)

K44 Use coloring theorems to prove the following about the vertex independence number of a graph. In each case n is the number of vertices.

(a) For any graph, $\alpha \geq n/(\Delta + 1)$.

(b) For a connected graph that is not complete and not an odd cycle graph, $\alpha \geq n/\Delta$.

(c) For a planar graph, $\alpha \geq n/4$.

K45 Looking back at the proof of the Six Color Theorem, it appears that the only property of planar graphs that was used was the fact that every planar graph contains at least one vertex of degree ≤ 5. This suggests trying a similar argument to prove that every graph that contains at least one vertex of degree ≤ 5 has a proper vertex coloring using at most six colors.

(a) But that isn't true. Find a counterexample.

(b) Where is the error in this "proof"?

K46 The graph below contains one uncolored vertex U. Without changing any of the existing colors, we would need a fifth color at U to color the graph properly. However find a color switch that changes the color at only one of the vertices adjacent to U. What does this allow you to do?

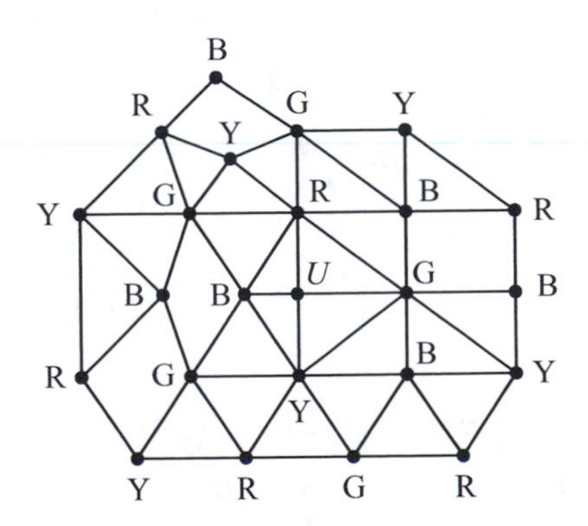

K47 Explain why the color switch in the preceding problem necessarily had to involve blue and green vertices.

Proof of the Four Color Theorem?

Attempting to prove the Four Color Theorem, we proceed as in the proof of the Five Color Theorem (problems K23–30), assuming now that every smaller planar graph has a proper vertex coloring that uses no more than four colors. Again, there must be a vertex U of degree ≤ 5 and we suppose that all other vertices have been colored properly using no more than four colors. We have to show that U can be colored without introducing a fifth color.

K48 If at most three colors occur at vertices adjacent to U, then there is no problem. Why?

If all four colors occur at vertices adjacent to U, we try to eliminate one of these by a color switch. One possibility is that U has degree 4, in which case we assume this arrangement of colors:

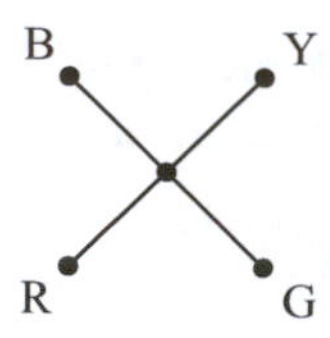

K49 Show that either a blue-green switch or a red-yellow switch reduces the number of colors that occur at vertices adjacent to U. If necessary, see the hint for problem K29.

If U has degree 5, then there are essentially two different possibilities for the arrangement of colors at vertices adjacent to U. In either case, one color occurs twice. The difference is in where the repeated color occurs.

K50 Show that in this case, either a red-yellow switch eliminates a color or else there is a blue-green switch that changes the blue vertex. In either case a color is eliminated from the neighbors of U.

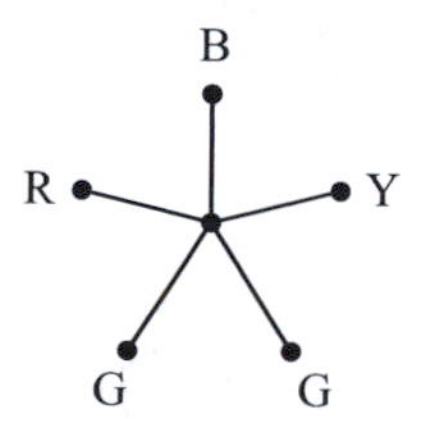

Finally, in the only remaining case, we assume this arrangement of colors:

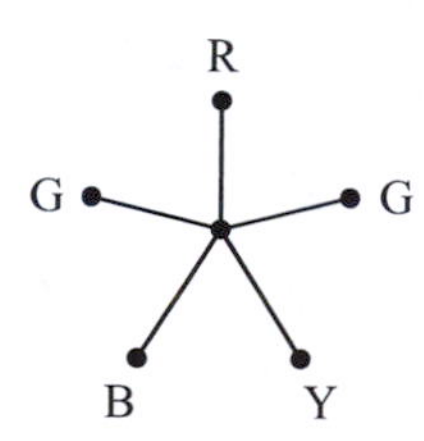

K51 (a) Show that either a red-blue switch eliminates a color or else there is a green-yellow switch that changes the green vertex at the left.

(b) Show that either a red-yellow switch eliminates a color or else there is a blue-green switch that changes the green vertex at the right.

Problem K51 suggests that there is always a way to eliminate one color at vertices adjacent to U: If a red-blue switch doesn't work and a red-yellow switch doesn't work, then both green vertices can be changed and we obtain this arrangement of colors:

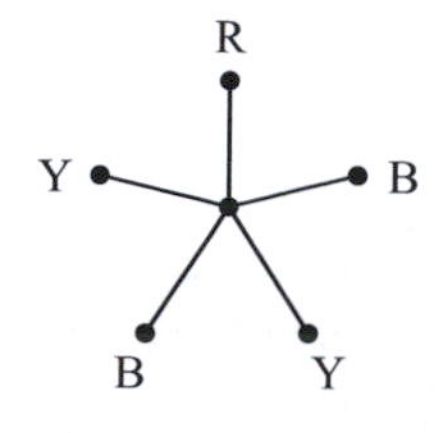

K52 The procedure given above seems to prove the Four Color Theorem. Try it on this graph to eliminate a color adjacent to the uncolored vertex. What goes wrong?

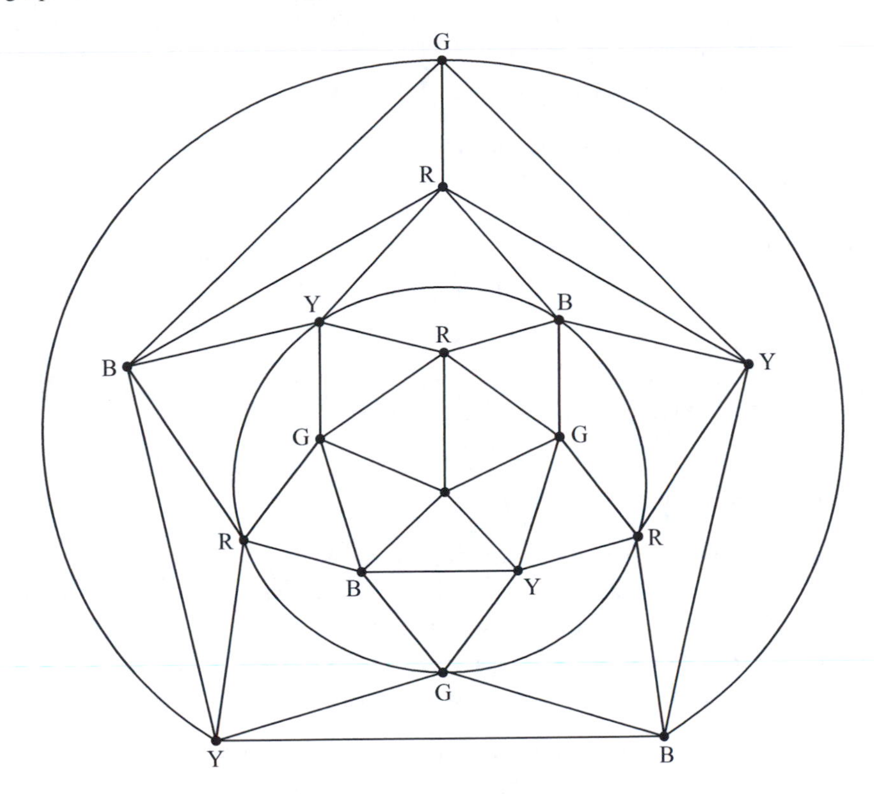

Historical note The argument presented in problems K48–K51 was originated by Alfred Kempe in 1879 and was considered to be a valid proof of the Four Color Theorem for 11 years. Then in 1890 Percy Heawood discovered an example similar to the one above that exposed the flaw in Kempe's method. Despite this, Kempe's ideas provided the basis for the eventual solution by Appel and Haken nearly a century later.

Proof of Brooks' Theorem for regular graphs

In the following problems we complete the proof of Brooks' Theorem (the second part of Vertex Coloring Theorem #2). The proof for the non-regular case appeared in problems K14 and 15, so we assume here that **G** is a regular graph. We assume also that **G** is connected.

K53 What does Brooks' Theorem say about 2-regular connected graphs? Verify that this is correct.

Having dispensed with the 2-regular case, assume now that **G** is a d-regular connected graph for some $d \geq 3$. We have to show that **G** has a proper vertex coloring using at most d colors unless **G** is complete. The concept of a **vertex cut set** becomes important here (see chapter J if necessary) because the proof requires different arguments for graphs with different minimal vertex cut sets. We begin with the easiest case.

K54 (a) Show that if **G** has a vertex cut set consisting of a single vertex A, then **G** is the union of two or more subgraphs $\mathbf{G}_1, \mathbf{G}_2, \ldots$ having only A in common.

(b) Show that each $\mathbf{G}_i$ is connected, non-regular, and has maximum degree d. What does this tell you about $\chi(\mathbf{G}_i)$?

(c) It follows that $\chi(\mathbf{G})$ is at most d. Why?

Next we assume that a minimal vertex cut set in **G** contains at least two vertices. If **G** is complete, there is nothing to prove, so we assume **G** is not complete. Remember that **G** is connected and d-regular with $d \geq 3$.

The next part of the argument requires three vertices A, B, and Z with the property that A and B are each adjacent to Z but nonadjacent to each other.

K55 Prove that **G** must contain three such vertices.

The idea is to color the vertices of **G** starting with A and B, assigning the same color to A and B, and ending with Z.

K56 (a) Suppose that the remaining vertices can be ordered in such a way that each vertex except Z is adjacent to some vertex that comes later in the ordering. Show that no more than d colors are needed. Why is there no problem at Z?

(b) Prove that such an ordering exists if $\{A, B\}$ is not a vertex cut set. If necessary, look back at the remarks following problem K14.

In the only remaining case, **G** has a minimal vertex cut set consisting of two nonadjacent vertices A and B. **G** splits into two or more components $\mathbf{H}_1, \mathbf{H}_2, \ldots$ when A, B and all edges at A and B are removed. From each $\mathbf{H}_i$, form a graph $\mathbf{G}_i$ by adding vertices A and B; all edges at A and B that go to vertices in $\mathbf{H}_i$; and an edge joining A and B. Any two of these graphs have in common only A, B and the edge joining them, and their union is the graph $\mathbf{G}^*$ consisting of **G** and the extra edge AB.

K57 Below are three examples of graphs with vertex cuts $\{A, B\}$. In each case draw the graphs $\mathbf{G}_1$ and $\mathbf{G}_2$. Which of the $\mathbf{G}_i$ are complete? Which are regular?

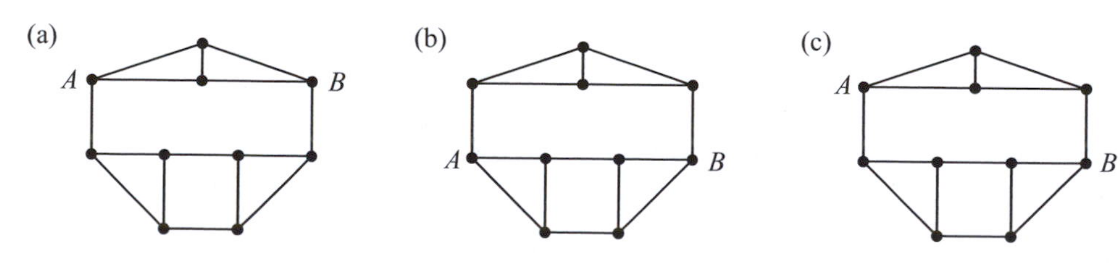

K58 Returning to the general situation, show that

(a) Each $\mathbf{G}_i$ is a connected graph with maximum degree d.

(b) If each $\mathbf{G}_i$ has a proper vertex coloring using at most d colors, then the same is true for $\mathbf{G}^*$, and then the same is true for **G**.

(c) Explain the role of the extra edge AB in the foregoing argument. What could go wrong if it were not included?

K59 Show that if all of the $\mathbf{G}_i$ are non-regular, then $\mathbf{G}$ has a proper vertex coloring using at most d colors. In which cases below are all $\mathbf{G}_i$ non-regular?

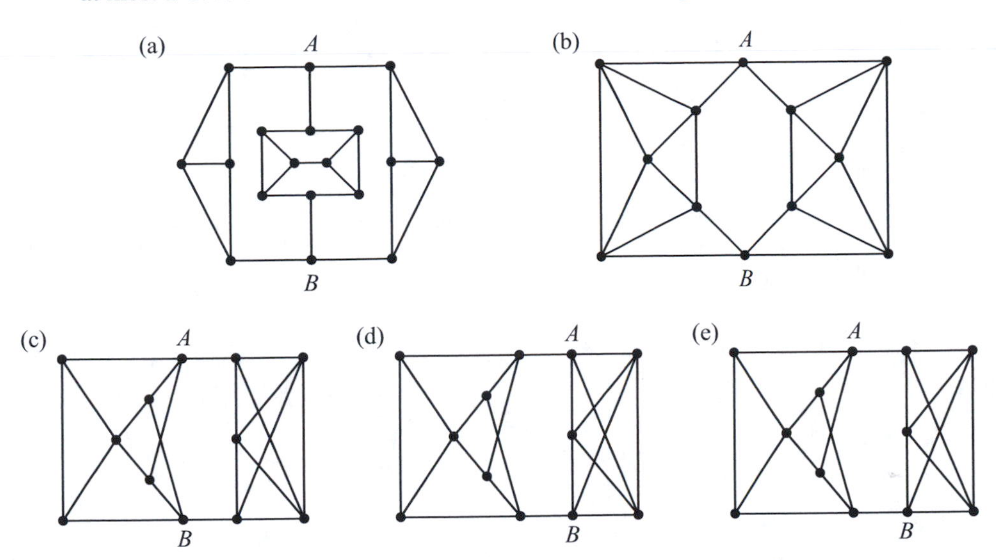

The remaining step in the proof consists of showing that there is always a choice of vertices A and B such that all $\mathbf{G}_i$ are non-regular. The idea for proving this is suggested by the graphs in problem K57 and the last three in K59.

K60 Suppose $\mathbf{G}_1$ is a regular graph. Prove that

(a) A and B are each adjacent to only one vertex outside of $\mathbf{G}_1$.

(b) There is only one other $\mathbf{G}_i$ graph: in other words, $\mathbf{G}^*$ is the union of $\mathbf{G}_1$ and $\mathbf{G}_2$.

(c) A and B are not both adjacent to the same vertex in $\mathbf{H}_2$.

(Hint for (b) and (c): Remember that $\{A, B\}$ is a minimal vertex cut set.)

Finally, we replace $\{A, B\}$ with the set $\{A, C\}$, where C is the vertex in $\mathbf{H}_2$ adjacent to B. This changes $\mathbf{G}_1$ and $\mathbf{G}_2$, adding C to $\mathbf{G}_1$ and removing B from $\mathbf{G}_2$. Edge AC replaces AB in both graphs.

K61 Show that the new graphs $\mathbf{G}_1$ and $\mathbf{G}_2$ are each non-regular. As in problem K59, that completes the proof of Brooks' Theorem.

L

Edge Coloring

In this chapter we consider the problem of coloring the edges of a graph.

Definition A *proper edge coloring* of a graph is the assignment of a color to each edge of the graph in such a way that no color occurs on edges that share a common endpoint.

Example

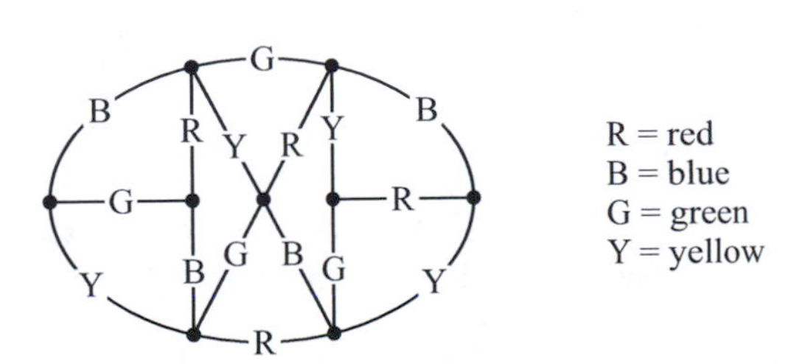

L1 (a) State this condition in another way by filling in the blank: In a proper edge coloring, the set of edges of any one color is _____.

(b) What must be true about the colors that occur on the edges at any vertex?

The Edge Coloring Number of a Graph

We are interested in coloring the edges of a given graph **G** properly using the smallest possible number of different colors. Such a coloring is called a *minimal proper edge coloring* of **G**. The *edge coloring number* χ' of **G** is the number of colors that occur in a minimal proper edge coloring.

L2 Is the coloring in the example above minimal? How do you know?

L3 What relationship exists between χ' and the maximum degree Δ in any graph?

L4 Find a minimal proper edge coloring for this graph. How does χ' compare with Δ?

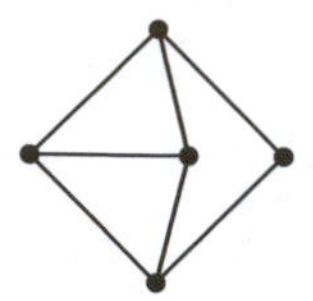

L5 Find a minimal proper edge coloring for the complete bipartite graph $\mathbf{K}_{3,4}$ by working on a 3-by-4 matrix. Assign a color to each cell. What property should the arrangement of colors in the matrix have so that the corresponding edge coloring is proper?

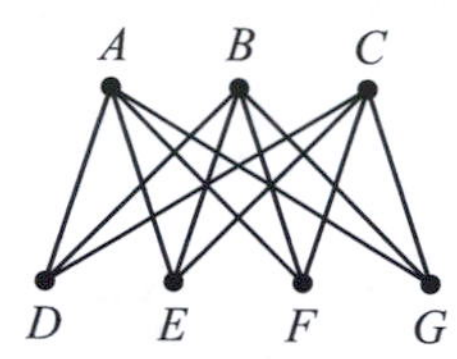

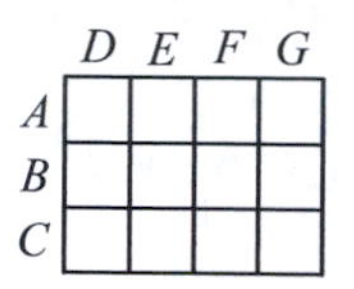

L6 Generalize the preceding problem to determine the edge coloring number of the complete bipartite graph $\mathbf{K}_{m,n}$. Assume that $m \leq n$.

L7 Find a minimal proper edge coloring for each graph below.

(a)

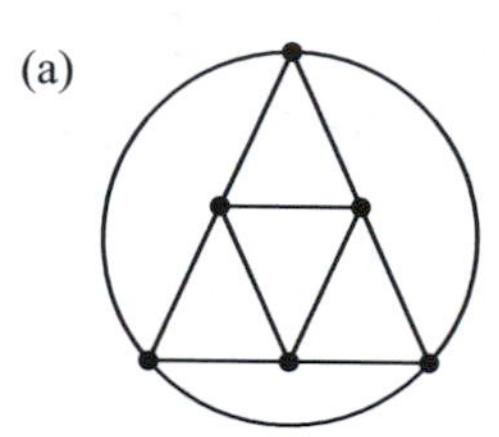

(b)

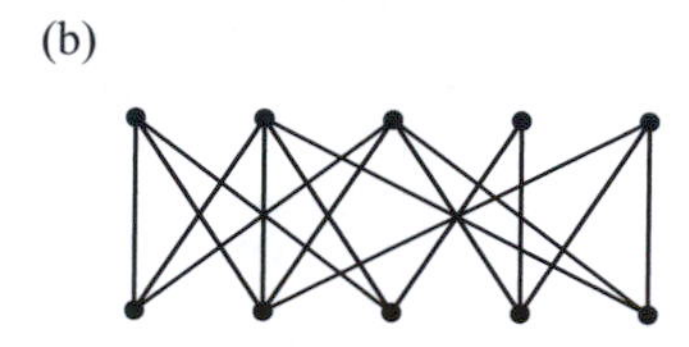

Suggestion for (b): The graph is bipartite, so work on a matrix as in problem L5. However some of the cells should be left empty.

Edge Coloring of Complete Graphs

L8 (a) Show that in any proper edge coloring of $\mathbf{K}_5$, no color can occur more than twice.

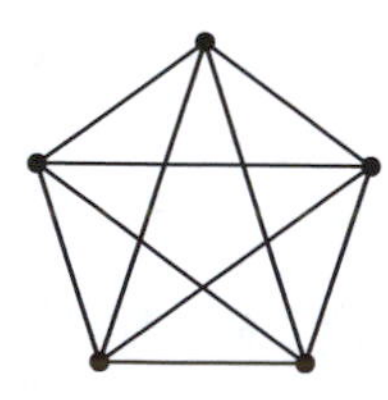

(b) The observation in (a) implies that four colors could not be enough for a proper edge coloring of $\mathbf{K}_5$. Why?

(c) Find a proper edge coloring of $\mathbf{K}_5$ using five colors. (Hint: Think about parallel edges.)

L9 Problem L8 suggests an edge version of Vertex Coloring Theorem #1. First fill in the blanks: In any graph with edge independence number α' and edge coloring number

χ', the number of edges of any particular color is at most _____, and therefore the total number of edges in the graph is at most _____.

This leads directly to a lower bound for χ':

Edge Coloring Theorem #1 *In any graph with e edges and edge independence number α', the edge coloring number χ' satisfies the inequality $\chi' \geq e/\alpha'$.*

L10 Use the inequality established above to show that in any complete graph $\mathbf{K}_n$ with an odd number of vertices $n \geq 3$, the edge coloring number is greater than or equal to n. (Suggestion: Let $n = 2m + 1$ and show that the number of edges is mn. What is α'?)

The edge coloring of $\mathbf{K}_5$ suggested by the hint in problem L8(c) can be generalized to $\mathbf{K}_n$ for any odd $n \geq 3$, producing a proper edge coloring of $\mathbf{K}_n$ using n colors.

L11 Try this for $\mathbf{K}_7$. Begin by drawing a regular septagon. Using problem L10, we conclude that $\chi' = n$ for any complete graph $\mathbf{K}_n$ with an odd number of vertices $n \geq 3$. Next we look at $\mathbf{K}_n$ when n is even.

L12 Find χ' for $\mathbf{K}_2$ and $\mathbf{K}_4$. How does χ' compare with n in these cases?

For even values of n, the edge coloring number of $\mathbf{K}_n$ is always $n - 1$.

L13 It is obvious that a proper edge coloring of $\mathbf{K}_n$ requires at least $n - 1$ colors. Why?

On the other hand, it is not so easy to see how to color the edges of $\mathbf{K}_n$ properly when n is even using only $n - 1$ colors. One way to do this is suggested by the next problem.

L14 Find a proper edge coloring of $\mathbf{K}_6$ using five colors by working with the diagram below and starting with an edge coloring of $\mathbf{K}_5$.

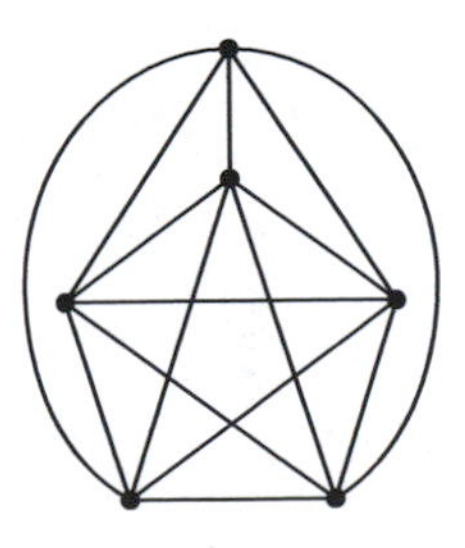

Summarizing, we have these results for the complete graph $\mathbf{K}_n$:

$$\chi' = n \quad \text{if } n \text{ is odd and} \geq 3$$
$$\chi' = n - 1 \quad \text{if } n \text{ is even.}$$

The next theorem shows that the edge coloring number of a graph, unlike the vertex coloring number, is limited to only two possibilities.

Edge Coloring Theorem #2 (the Vizing-Gupta Theorem) *In any graph with maximum degree Δ, χ' is either Δ or $\Delta + 1$.*

To say this another way, every graph falls into one of two categories:
Vizing class 1, consisting of graphs with $\chi' = \Delta$, or
Vizing class 2, consisting of graphs with $\chi' = \Delta + 1$.

L15 Determine the Vizing class of each of these graphs.

(a) The cycle graph $\mathbf{C}_n$

(b) The complete graph $\mathbf{K}_n$

(c) The complete bipartite graph $\mathbf{K}_{m,n}$

(d) The graph preceding problem L1

(e) The graph in problem L4

(f) The graphs in problem L7

A proof of Edge Coloring Theorem #2 will be given in problems L40–L48.

Edge Coloring of Bipartite Graphs

For a bipartite graph, there is no mystery about the edge coloring number.

Edge Coloring Theorem #3 (König's Coloring Theorem) *In any bipartite graph with maximum degree* Δ, $\chi' = \Delta$.

In other words, every bipartite graph is in Vizing class 1.

L16 According to this theorem, the graph below has a proper edge coloring using four colors. Find it.

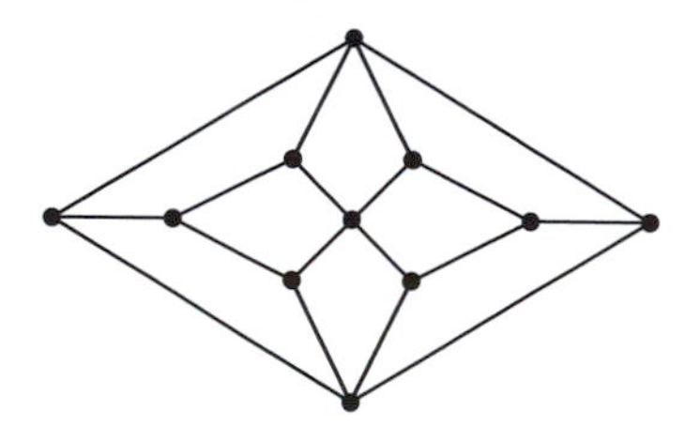

Edge Color Switch

König's Coloring Theorem can be proved by a method similar to the vertex color switch used in the proof of the Five Color Theorem in Chapter K. In an *edge color switch*, two colors are interchanged on a set of edges without creating any edges of the same color with a common endpoint.

L17 In the graph below only the center edge UV is uncolored. Each of the four colors occurs at one or both of the endpoints of this edge, so it appears that a fifth color might be needed. However no blue edge occurs at U and no green edge occurs at V, so look for a blue-green color switch that eliminates a color at one endpoint of UV without introducing it at the other. What does this allow you to do?

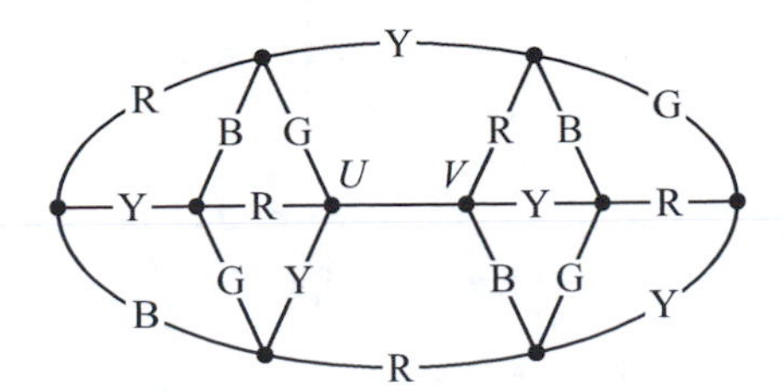

L18 Again, in the graph below, we want to eliminate a color at one endpoint of the uncolored edge UV without introducing it at the other.

(a) Try a green-white switch. What goes wrong?

(b) Find a color switch that makes a color available for UV.

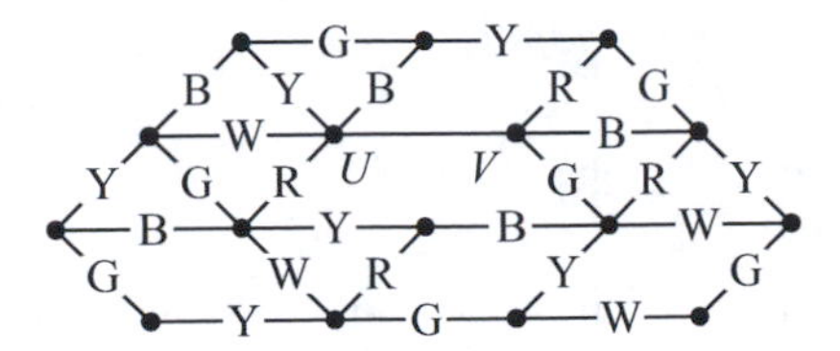

Although the graphs in problems **L17** and **L18** are not bipartite, they illustrate the method that will be used to prove König's Coloring Theorem. The idea is to use a color switch to make a color available for an uncolored edge.

Proof of Edge Coloring Theorem #3

Let **G** be a bipartite graph with maximum degree Δ. We claim that it is possible to color the edges one at a time, in any order, using a fixed set of Δ colors.

L19 Let UV be any edge encountered in the coloring process. At each endpoint U and V, there must be at least one color that has not yet been used on any edge at that vertex. Why?

If the same color is absent at both U and V, then this color can be assigned to UV and we can move on to the next edge. In any other case, we can assume that a blue edge occurs at U but not at V and a green edge occurs at V but not at U.

L20 The situation described above occurs in this bipartite graph.

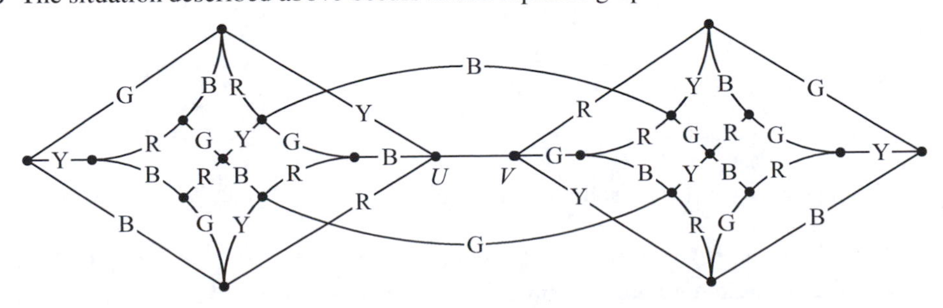

Find a color switch that eliminates blue at U without introducing it at V.

Returning to the general case, we claim that there is always a blue-green color switch that eliminates blue at U without introducing it at V.

L21 What is the only thing that might go wrong? Show that this cannot happen in a bipartite graph. (Hint: Think about the lengths of paths and cycles.)

Problem L21 shows that it is always possible to make a color available for an uncolored edge in a bipartite graph without using more than Δ colors. That completes the proof of Edge Coloring Theorem #3.

Application of Edge Coloring: the Scheduling Problem

L22 The edges of the graph below indicate interviews that have to be scheduled between job applicants $A_1, A_2, \ldots$ and employers $E_1, E_2, \ldots$. Each interview occupies a 15-minute time period and no person can be in two places at the same time. Design a schedule that allows all interviews to be completed within one hour. Do this by placing numbers 1, 2, 3, 4 in the appropriate cells of the matrix. However before starting, explain how König's Coloring Theorem guarantees that it can be done.

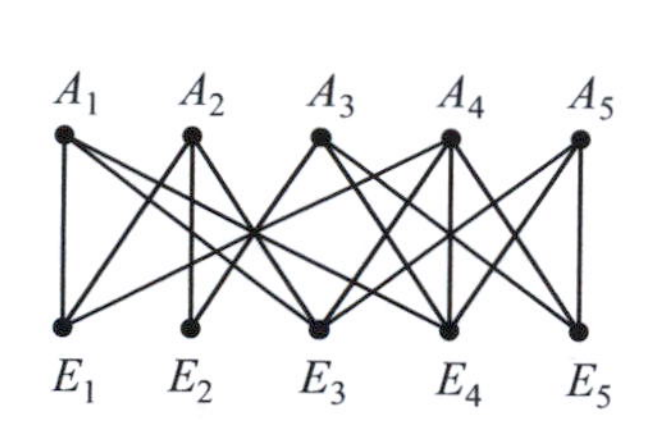

	E_1	E_2	E_3	E_4	E_5
A_1					
A_2					
A_3					
A_4					
A_5					

1 1:00 to 1:15
2 1:15 to 1:30
3 1:30 to 1:45
4 1:45 to 2:00

More Problems

L23 Find a minimal proper edge coloring for each graph below.

(a)

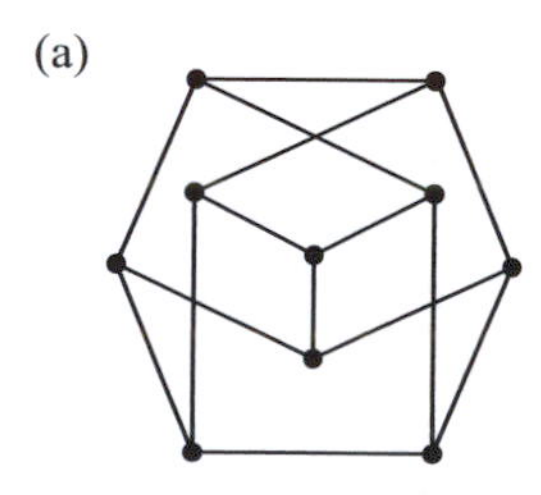

(b)

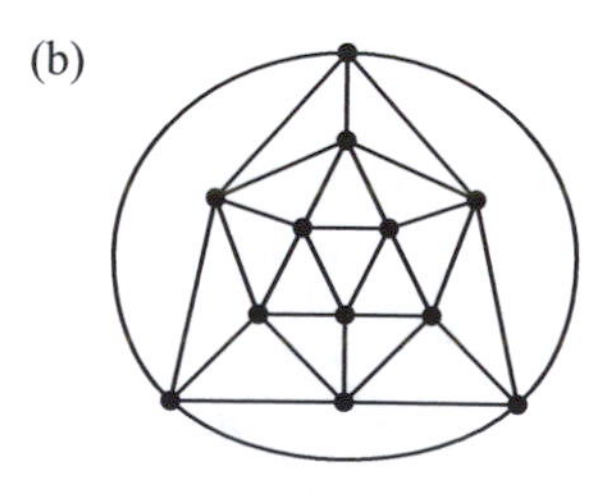

Suggestion for (b): Start by finding a perfect matching. How does this help?

L24 To color the edges of this graph, start by alternating two colors around a Hamilton cycle. If the remaining edges require three more colors, go back and try a different Hamilton cycle.

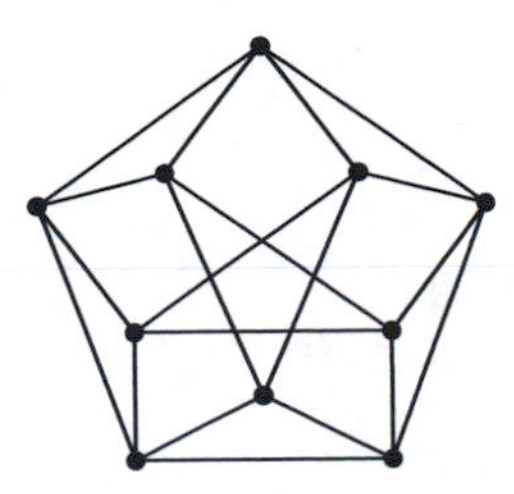

L25 Let **H** be a subgraph of **G**. Is it necessarily true that $\chi'(\mathbf{H}) \leq \chi'(\mathbf{G})$? Explain.

L26 (a) Does Edge Coloring Theorem #1 give any useful information about this graph?

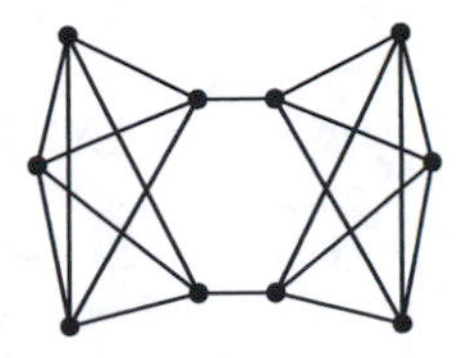

(b) Now see what the same theorem says about this graph.

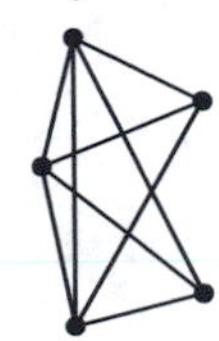

What does this tell you about the edge coloring number of the graph in (a)?

L27 (a) Find an arrangement of the numbers 1, 2, 3, 4, 5 in a 5-by-5 matrix satisfying the following conditions:

(1) Each number appears exactly once in each row;

(2) Each number appears exactly once in each column;

(3) The matrix is symmetric: the number in row i and column j is the same as the number in row j and column i.

(b) Use this matrix to determine minimal proper edge colorings of $\mathbf{K}_5$ and $\mathbf{K}_6$.

L28 Let **G** be a graph with e edges, maximum degree Δ, and edge independence number α' satisfying $\alpha' < e/\Delta$. Prove that **G** must be in Vizing class 2.

L29 Use the result of the preceding problem to show that the graph below is in Vizing class 2.

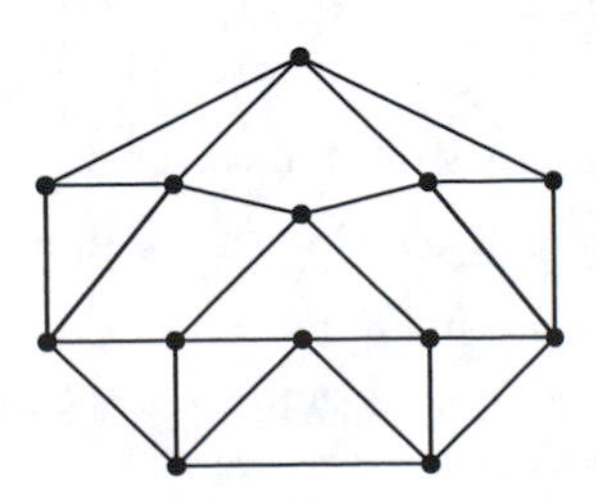

L30 Prove that every regular graph with an odd number of vertices (at least 3) is in Vizing class 2.

L31 (a) Find a multigraph with $\chi' = \Delta + 2$.

(b) Find a multigraph with $\chi' = \Delta + 100$.

L32 Does König's Coloring Theorem generalize to bipartite multigraphs? Look back at the proof in problems L19 to L21 and see if the argument applies equally well when multiple edges are allowed.

L33 Find a color switch in this graph that makes a color available for edge UV.

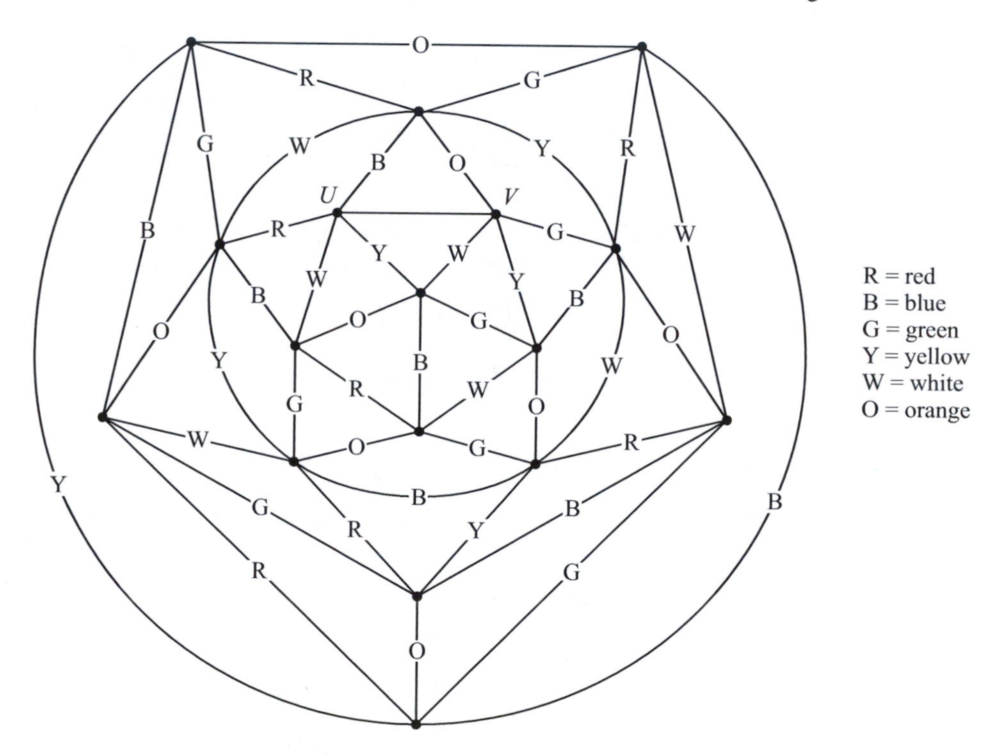

L34 What's wrong with this "proof" of König's Coloring Theorem? If **G** is any bipartite graph, add edges until you get a complete bipartite graph **G**$'$. It's easy to prove that $\chi' = \Delta$ for a complete bipartite graph, and since **G** is a subgraph of **G**$'$, it follows that $\chi'(\mathbf{G}) \leq \chi'(\mathbf{G}') = \Delta$.

L35 Find a way to insert letters A, B, C, D, E, F, G, H into the empty section of the first matrix below so that

(1) each letter appears exactly once in each row, and

(2) no letter appears more than once in any column.

Do this by first inserting column numbers 4, 5, 6, 7, 8 into the second matrix. What condition must be satisfied by the numbers in each row and column? What does this have to do with König's Coloring Theorem?

A	D	G
E	H	C
B	F	D
G	B	E
H	C	A

A	B	C	D	E	F	G	H
1			2			3	
		3		1			2
	1		3		2		
	2			3		1	
3		2					1

L36 (a) Suppose that **G** is a 3-regular graph that contains a Hamilton cycle. Show that **G** has a proper edge coloring using three colors.

(b) Prove that this graph does not contain a Hamilton cycle.

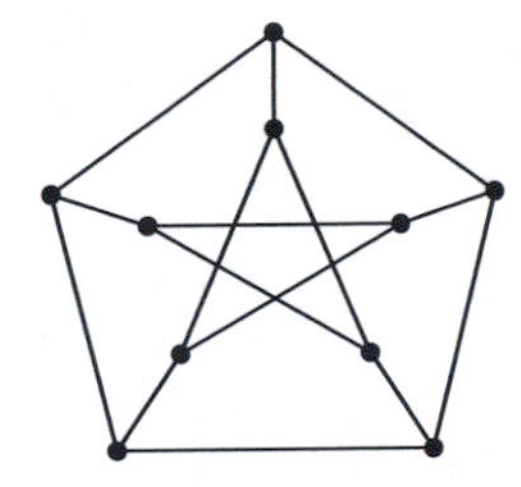

L37 Find all connected graphs in which the vertex coloring number is greater than the edge coloring number. (Hint: Think about what each of these would have to be in relation to Δ.)

L38 Use coloring theorems to prove the following about the edge independence number of a graph. In each case e is the number of edges.

(a) For any graph, $\alpha' \geq e/(\Delta + 1)$.

(b) For a bipartite graph with $\Delta \geq 1$, $\alpha' \geq e/\Delta$.

L39 Use König's Coloring Theorem to show that every d-regular bipartite graph with $d \geq 1$ contains a perfect matching. (This statement appeared earlier as the corollary to Independence Theorem #3. Since König's Coloring Theorem is easier to prove than Independence Theorem #3, this represents an easier way to prove the corollary.)

Proof of Edge Coloring Theorem #2

Let **G** be a graph with maximum degree Δ. We will show that it is possible to color the edges one at a time, in any order, using a fixed set of $\Delta + 1$ colors. In particular, we show that at any stage the coloring can be extended to one more edge, possibly by changing some of the previously assigned colors, until all edges have been colored properly.

L40 In this graph there is one uncolored edge, indicated by the symbol U.

(a) Find a recoloring of the three edges at vertex A so that all edges of the graph are colored properly using only the colors indicated at the right.

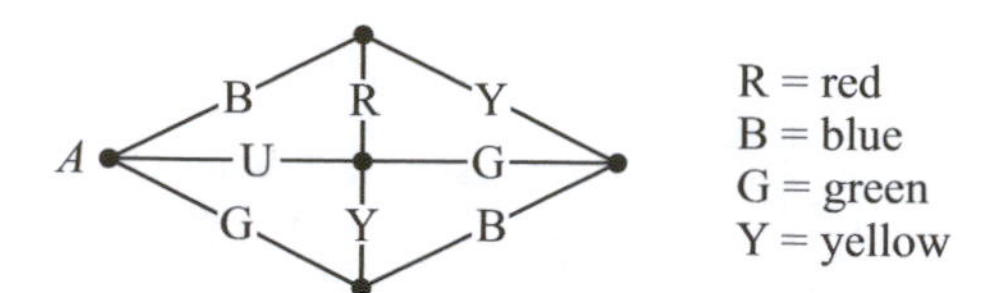

(b) What does this directed path have to do with the recoloring obtained above?

L41 Find four different simple directed paths from U to Y in this directed graph and use each one to determine a recoloring of the edges at vertex A in the graph at the right so that U is eliminated and all edges are colored.

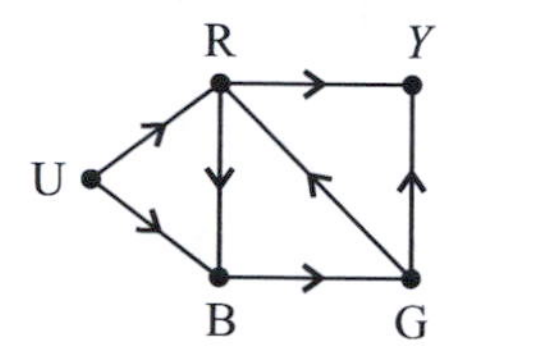

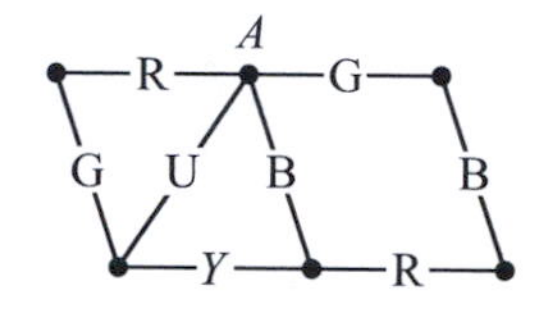

Problem **L41** shows how a directed graph can be used to determine a recoloring of the edges at a given vertex of a graph. In general, if AB is an uncolored edge of a partially edge-colored graph **G**, we construct a digraph **D** as follows: Vertices of **D** are labeled with colors that occur on edges of **G**, along with the symbol U. At the vertex labeled U, only outgoing edges exist and these go to colors that are absent at vertex B. An edge goes from X to Z in **D** whenever color X occurs on some edge at A and color Z is absent at the other (non-A) endpoint of that edge.

in **G**: A •—X—•< (no Z occurs) in **D**: X •—>—• Z

Because of its dependence on vertex A, **D** will be referred to as the *A-based digraph* associated with **G**.

L42 (a) Construct the A-based digraph associated with this graph. There should be six vertices and eight edges.

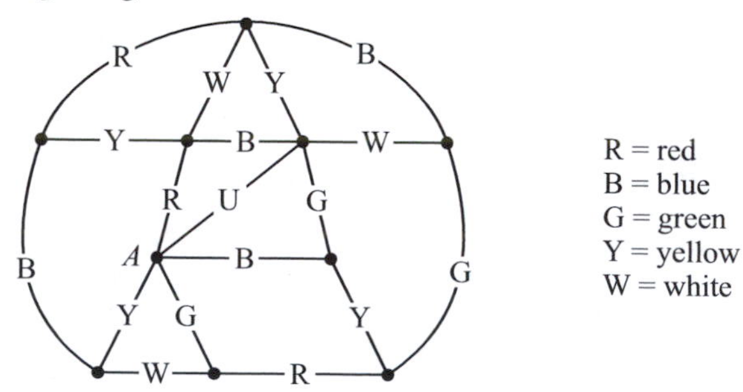

(b) Use the digraph obtained above to eliminate U from the edges at A.

L43 What goes wrong when you try to apply the method of the preceding problem to this graph?

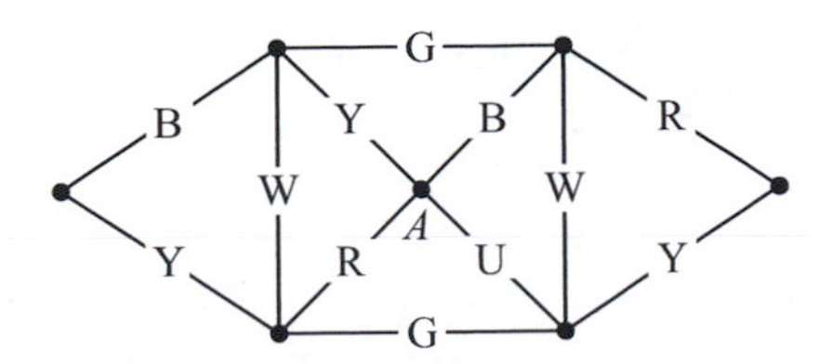

The success of the recoloring process in problems L41 and L42 depended on the existence of a path starting at U in the A-based digraph and ending at a color not occurring at A. That color was then inserted on one edge at A with all other colors along the path shifting to other edges and U being eliminated. The remaining problem, then, is to find a suitable recoloring of edges when all colors that are reachable from U already appear on edges at A.

L44 Assuming that $\Delta + 1$ colors are available for coloring edges of **G**, show that if color X occurs at A, then at least one outgoing edge exists at vertex X in the A-based digraph. What can you say about the number of outgoing edges at U?

Problem L44 shows that a path starting at U will eventually either reach a color not occurring at A or else it will reach a point where an outgoing edge leads back to a color that occurred earlier on the path.

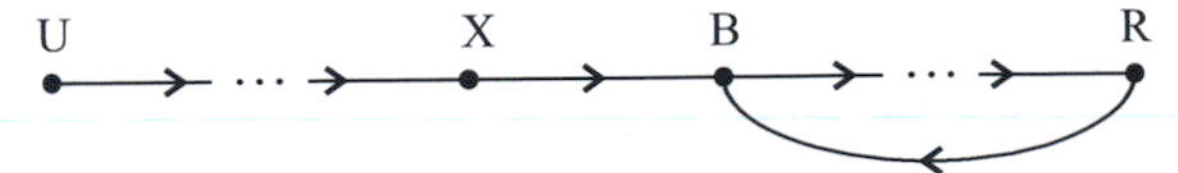

Here the outgoing edge at R (red) leads back to B (blue), and the vertex immediately preceding the first occurrence of B has the label X. We allow the possibility that X $=$ U.

Among the edges at A, let AC be the X-colored edge, let AD be the blue edge, and let AE be the red edge.

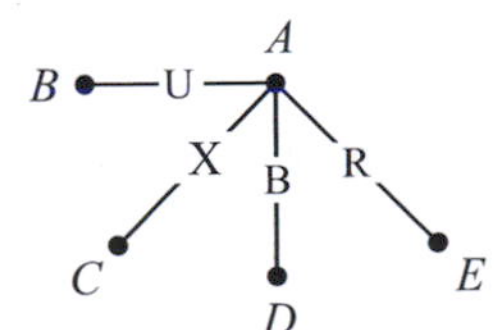

L45 No blue edges occur at vertices C and E. How do we know this?

The path from U to X determines a recoloring of edges at A that shifts U to edge AC. In other words, AC becomes the uncolored edge. AD remains blue, AE remains red, and blue remains absent at both C and E.

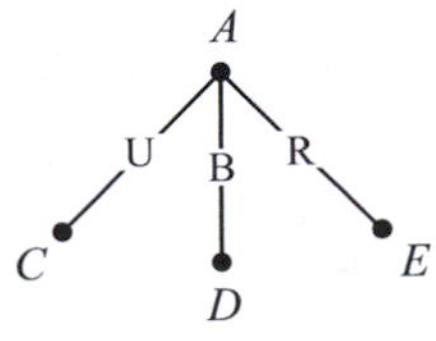

Assuming that green is one of the colors that is absent at A, we consider whether there is a blue-green color switch that eliminates a color at one endpoint of AC without introducing it at the other.

L46 (a) Why would we want to do that?

(b) Show that such a color switch exists unless **G** contains a blue-green path starting at A and ending at C.

In view of this, it remains to consider what happens when such a path **P** exists in **G**. In that case we continue to shift U according to the path in **D** until AE becomes the uncolored edge.

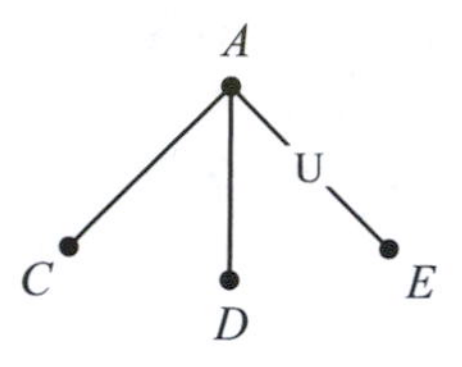

L47 (a) Which edge is now blue? How does that affect **P**?

(b) There is still a blue-green path starting at A, but now where does it end?

(c) Is blue still absent at E?

L48 Complete the proof by finding a color switch that eliminates a color at one endpoint of AE without introducing it at the other. What does that allow you to do?

Matching Theory for Bipartite Graphs

This chapter revisits the topic of independent vertex and edge sets and covering sets in a graph. We begin with some unfinished business from chapter I, the proof of the König–Egervary Theorem (Independence Theorem #3).

It will be helpful to review some definitions from chapter I.

An *independent vertex set* in a graph is a set of vertices that does not include any adjacent vertices. An *independent edge set* is a set of edges that does not include any edges that share an endpoint.

A *covering edge set* in a graph is a set of edges with the property that every vertex in the graph is an endpoint of at least one edge in the set.

A *covering vertex set* in a graph is a set of vertices with the property that every edge in the graph has at least one endpoint in the set.

The *independence numbers* for a given graph are

$$\alpha = \text{the number of vertices in a maximal independent vertex set,}$$

$$\alpha' = \text{the number of edges in a maximal independent edge set.}$$

The *covering numbers* for a given graph are

$$\beta = \text{the number of vertices in a minimal covering vertex set,}$$

$$\beta' = \text{the number of edges in a minimal covering edge set.}$$

M1 Find maximal independent vertex and edge sets and minimal covering vertex and edge sets in this graph. Use the two covering theorems in chapter I. If you don't find five independent vertices, look again.

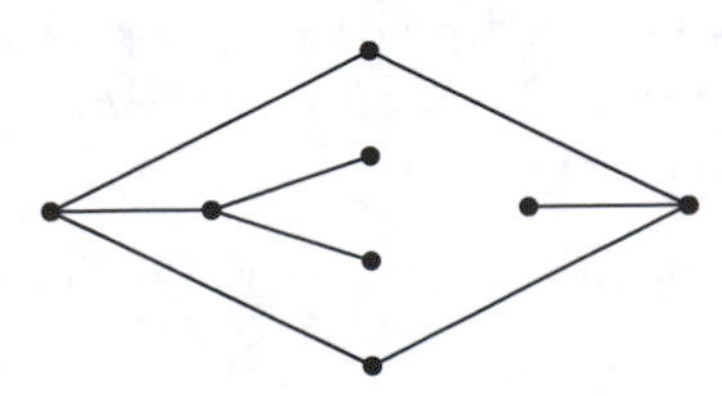

Also recall these:

Independence Theorem #2 (proved in chapter I) *In any graph with n vertices, $\alpha + \alpha'$ is always less than or equal to n.*

Independence Theorem #3 (not yet proved) *In any bipartite graph with n vertices, $\alpha + \alpha = n$.*

Independence Theorem #3 is one form of the König–Egervary Theorem. Other forms can be derived using these relations established in chapter I:

(1) In any graph with n vertices, $\alpha + \beta = n$.

(2) In any graph with n vertices and $\delta \geq 1$, $\alpha' + \beta' = n$.

M2 State an equivalent form of the König–Egervary Theorem

(a) In terms of α' and β,

(b) In terms of α and β',

(c) In terms of β and β'.

Assume $\delta \geq 1$ for (b) and (c).

M3 Recall that a *matching* in a graph is any independent edge set. Restate your answer to problem M2 (a) in words, using the term "matching."

M4 (a) Show that this graph is bipartite.

(b) Find four independent edges.

(c) Find a covering vertex set consisting of four vertices.

(d) Use your answer to (c) to find a set of seven independent vertices.

(e) Explain how you know that the sets found in (b) and (d) are each maximal.

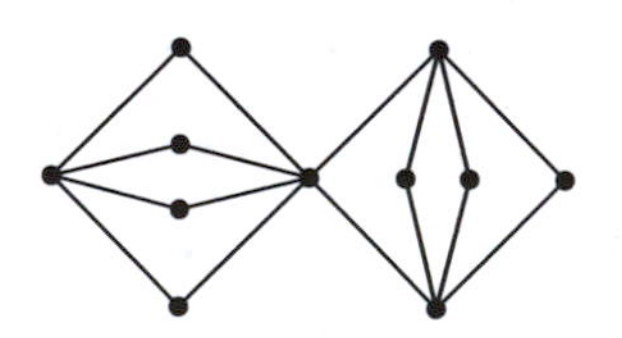

Often, as in the example above, it is possible to find an independent edge set in a graph and a covering vertex containing an equal number of members. Then the edge set must be maximal and the vertex set minimal. We can conclude this without looking further at the graph. Also, the graph does not have to be bipartite for this conclusion to be valid.

The Max/Min Principle

In any graph, if there is a set of k independent edges, for some number k, and a covering vertex set consisting of k vertices, then $\alpha' = \beta = k$. In other words, the edge set is maximal and the vertex set is minimal.

M5 Use the max/min principle to show that the covering vertex set found in problem M4 (c) is minimal.

M6 Prove the max/min principle by the following steps.

(a) Show that $\alpha' \geq k$ and $\beta \leq k$.

(b) Show that $\alpha \geq n - k$.

(c) It follows that equality must hold in both (a) and (b). How do we know this?

Proof of the König–Egervary Theorem

We now begin a series of problems that lead up to the proof of the König–Egervary Theorem. We will show that $\alpha + \alpha' \geq n$ in any bipartite graph with n vertices. Since we also know that $\alpha + \alpha' \leq n$ in any graph (Independence Theorem #2), it will follow that $\alpha + \alpha' = n$.

The proof that $\alpha + \alpha' \geq n$ will be accomplished by starting with a maximal independent edge set and using it to produce a sufficiently large independent vertex set.

M7 In this graph, $n = 11$ and $\alpha' = 4$. To show that $\alpha + \alpha' = 11$, we need to find seven independent vertices. Try to do this by trial and error.

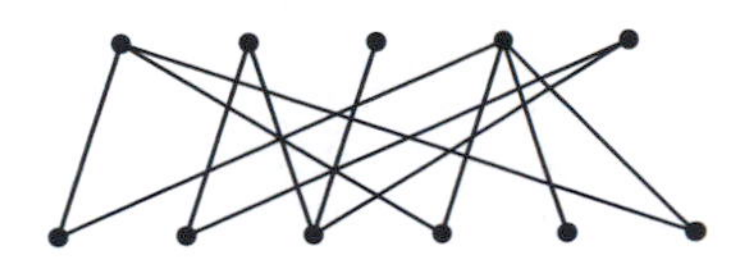

Returning to the general case, we will show how the independent vertex set we looked for in problem M7 can be constructed systematically by a process that applies to any bipartite graph. The first step is the construction of a directed graph, or digraph, in which each vertex is assigned one of three possible colors.

The Colored Digraph Construction

Let **G** be any bipartite graph and let **M** be any matching (an independent edge set) in **G**. We assume that the vertices of **G** are arranged in two rows, one above the other, with each edge having one vertex in each row.

The *colored digraph* **D** corresponding to **M** is constructed by the following rules.

Directions of edges Assign a direction to each edge of **G**. Edges of **M** are directed upward. All other edges are directed downward.

Example

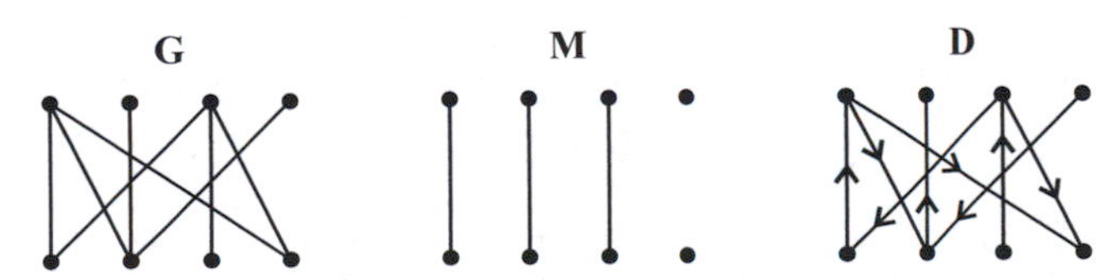

This makes **G** into a digraph. Notice that every directed path in this digraph alternates between upward and downward edges, and therefore it alternates between edges that are in **M** and edges that are not in **M**.

In what follows, we refer to endpoints of edges in **M** as *covered vertices*. All other vertices are *uncovered*.

Coloring of vertices:

Covered vertices in the top row are **blue**;
Covered vertices in the bottom row are **red**;
All uncovered vertices are **green**.

Example

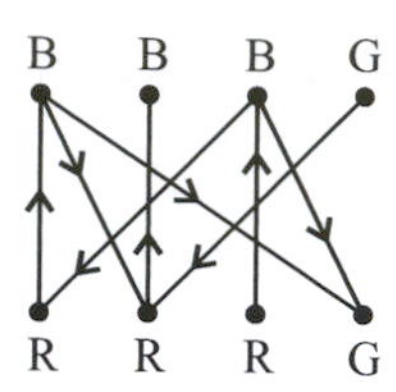

A vertex set **S** is then constructed from **D** by the following recipe:
S contains
all green vertices;
all blue vertices that are reachable in **D** from green vertices;
all red vertices that are not reachable in **D** from green vertices.

Example

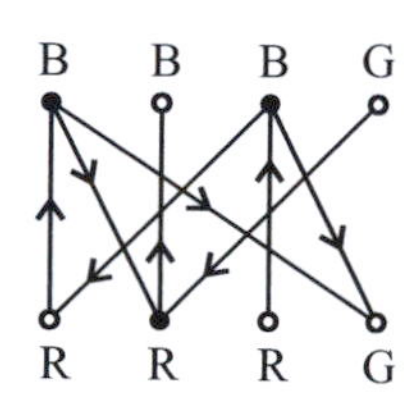

o indicates vertices in **S**

Notice that in this example **M** is a maximal matching and the resulting vertex set **S** is independent.

M8 (a) Show that this matching is non-maximal in the given graph.

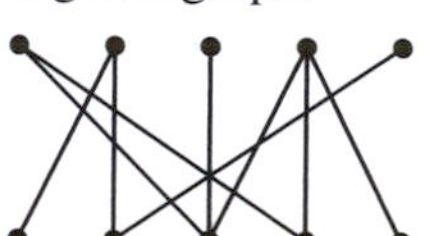

(b) Construct the corresponding colored digraph and find the vertex set **S**. Is **S** independent?

We will see that the vertex set **S** constructed from a given matching is independent if and only if the matching is maximal. Furthermore, whenever the matching is non-maximal there is a simple algorithm that increases it to a larger matching. The process is illustrated in the next problem.

M9 In the colored digraph from problem M8, find a path **P** from one green vertex to another green vertex and change the matching **M** as follows:
All edges of **P** that are in **M** are removed from **M**.
All edges of **P** that are not in **M** are added to **M**.
What does this change accomplish?

Returning to the general case, we have this algorithm.

Matching Extension Algorithm

Let **P** be a path from one green vertex to another green vertex in the colored digraph constructed from a given matching **M** in a bipartite graph. Then **M** can be increased to a larger matching by removing all edges of **P** from **M** and adding all edges of **P** that are not in **M**.

M10 Explain why this works. Show that the new edge set is still a matching and that it contains exactly one more edge than **M**.

The matching extension algorithm shows how a non-maximal matching can be increased whenever a green-to-green path exists in the corresponding colored digraph. However, does such a path necessarily exist? The answer is provided by the following theorem.

The Colored Digraph Theorem *These three conditions are equivalent; if any one of them holds for a given matching* **M** *in a bipartite graph, then all three conditions hold.*

(1) **M** *is maximal;*

(2) **D** *contains no path with two green endpoints;*

(3) **S** *is an independent set.*

This important result, which will be proved in problems M14–16, has both practical and theoretical consequences. On the practical side, it shows that a non-maximal matching in a bipartite graph can always be increased to a larger matching by the matching extension algorithm, since it implies that the corresponding **D** contains a green-to-green path. On the theoretical side, it allows us to prove the König–Egervary Theorem. How this is done is explained below.

Recall the definition of the set **S**. It contains

all green vertices;
all blue vertices that are reachable in **D** from green vertices;
all red vertices that are not reachable in **D** from green vertices.

M11 Show that each edge in **M** has exactly one endpoint in **S**. Specifically, show that if the red endpoint of the edge is reachable from a green vertex then so is the blue one, and conversely. (Hint: How many incoming edges occur at a blue vertex?)

M12 Use the preceding problem to show that if the entire bipartite graph contains n vertices and **M** contains m edges, then **S** contains exactly $n - m$ vertices.

M13 Show how the König–Egervary Theorem follows from the Colored Digraph Theorem. Use the Colored Digraph Theorem to show that if **M** is a maximal matching, then **S** is an independent set containing $n - \alpha'$ vertices.

Proof of the Colored Digraph Theorem

Finally we complete the reasoning by proving the Colored Digraph Theorem.

M14 (1) *implies* (2): If **M** is maximal, then **D** contains no green-to-green path. Why is this true?

M15 (2) *implies* (3): If **D** contains no green-to-green path, then **S** is independent. Consider each possibility for adjacent vertices in **S** and show that none of them can occur. See the hints below only if necessary.

(a) green and green
(b) green and red
(c) blue and green
(d) blue and red joined by an upward edge
(e) blue and red joined by a downward edge

Hints: (b) Which way would the edge go? Remember that the red vertex is in **S**. (c) There would be a green-to-green path. (d) See problem M11. (e) The red vertex would be reachable from a green vertex.

M16 (3) *implies* (1): If **S** is independent, then **M** is maximal.

(a) Using the notation of problem M12, show that $\alpha \geq n - m$ and $\alpha' \geq m$.
(b) Equality must hold in (a). Why?
(c) How does this show that **M** is maximal?

The proof of the Colored Digraph Theorem is now complete. Consequently so is that of the König–Egervary Theorem.

Matrix Interpretation of the König–Egervary Theorem

Next we turn our attention to a topic that might appear at first to be unrelated to what we have been talking about but which, as we will see, is just another way of looking at the same thing.

Below is a 4-by-4 array of cells, each of which is either open (blank) or closed (indicated by ⊠). We will call this a *matrix of open and closed cells*. Alternatively, the same information could be represented by a matrix of zeros and ones, with the ones representing open cells.

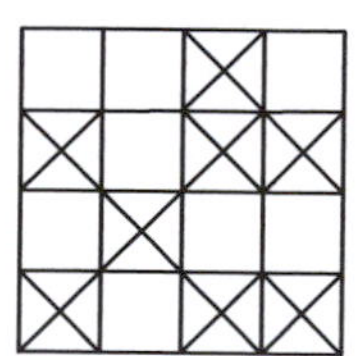

$$\begin{array}{cccc} 1 & 1 & 0 & 1 \\ 0 & 1 & 0 & 0 \\ 1 & 0 & 1 & 1 \\ 0 & 1 & 0 & 0 \end{array}$$

Definitions A *line* in a matrix is either a row (horizontal) or a column (vertical). A set of cells is *independent* if no two cells are in the same line.

M17 Find a maximal independent set of open cells in the matrix above. How can you be certain that there is no larger independent set?

M18 Do the same for this matrix.

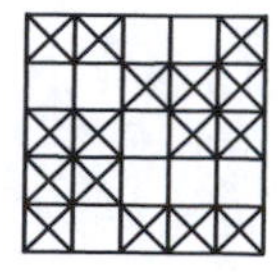

Actually, this is nothing new. A matrix of open and closed cells is equivalent to a bipartite graph if we construct a correspondence in the right way:

M19 Exactly what does this mean? Draw a bipartite graph that corresponds to the matrix in problem M17.

M20 What does an edge correspond to in a matrix of open and closed cells? What does a vertex correspond to?

M21 What does an independent edge set correspond to in a matrix of open and closed cells?

M22 A covering vertex set in a bipartite graph corresponds to a set of lines having a special property in a matrix of open and closed cells. What is that property?

M23 Restate the König–Egervary Theorem (the form that says $\alpha' = \beta$) in terms of a matrix of open and closed cells, using the terms "independent open cells" and "covering line set". What is a covering line set?

M24 Find a covering line set consisting of four lines in this matrix.

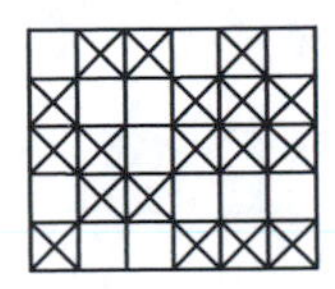

M25 State a max/min principle for independent open cells and covering line sets in a matrix. How does this apply to the matrix in problem M24?

According to the preceding problem, the matrix in problem M24 does not contain five independent open cells. Another way to see this is to notice that three of the rows (2, 3, and 5) contain open cells in only two columns (2 and 3).

M26 How does this show that the matrix does not contain five independent open cells? (Hint: Each row would have to include one of these cells.)

M27 If an m-by-n matrix of open and closed cells contains m independent open cells, then within any set of k rows (for any $k \leq m$), open cells occur in at least k different columns. Explain why this is true. Be careful to notice the difference between open cells and independent open cells.

M28 Complete this statement: If an m-by-n matrix of open and closed cells contains a set of k rows (for some $k \leq m$) in which open cells occur in fewer than k different columns, then ______________________________.

M29 We have seen how problem M28 applies to the matrix in problem M24. How does it apply to this matrix?

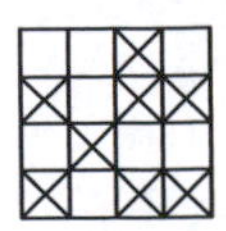

Hall's Theorem and Its Consequences

We continue to consider whether a given matrix contains a set of independent open cells that includes one cell in each row. Problem M27 gives a necessary condition for this to occur. Now we will see that this same condition is also sufficient.

Hall's Theorem *An m-by-n matrix of open and closed cells contains m independent open cells if and only if it has the following property for each $k = 1, 2, 3, \ldots, m$.*

Hall's condition: Within any set of k rows, open cells occur in at least k different columns.

It remains for us to prove that if Hall's condition is satisfied then the matrix contains m independent open cells. We do this by contradiction, using the matrix form of the König–Egervary Theorem. Suppose that the matrix does not contain m independent open cells.

M30 Consider a minimal covering line set in this matrix. Let r be the number of rows and c the number of columns in this line set. Show that the sum $r + c$ must be strictly less than m.

M31 Let k represent the number of rows in the matrix that are not in the line set of the preceding problem.

(a) These k rows contain open cells in at most c different columns. How do we know this?

(b) Why is that a contradiction?

That proves Hall's Theorem. In a theoretical sense, it solves the problem of whether a given m-by-n matrix contains m independent open cells. In a practical sense, however, it is not very convenient to apply.

M32 For a matrix with 20 rows, proving the existence of 20 independent open cells by Hall's Theorem would require checking more than a million things. Exactly what do we mean by this?

In practice, therefore, it is usually more efficient to take a direct approach to the existence of independent open cells in all rows by constructing a maximal matching in the corresponding bipartite graph. Sometimes, however, there is an easy way to predict that these cells exist.

Corollary 1 of Hall's Theorem *In an m-by-n matrix of open and closed cells, let h represent the maximum number of open cells in any column. If each row contains at least h open cells, then the matrix contains m independent open cells.*

M33 Prove this by answering the following:

(a) For any set of k rows, what can we say about the total number of open cells in these rows?

(b) How many of these open cells can occur in any one column?

(c) What does that tell you about the number of columns in which these open cells occur?

M34 For each matrix below, decide whether corollary 1 applies. In each case, does the matrix contain four independent open cells?

(a)

	X			X
X			X	
		X		
	X			X

(b)

X		X		X	
		X	X		X
	X			X	X
X	X		X		

(c)

	X		X	
X		X		
		X		X
X	X			X

Finally, we establish a special case in which the condition in corollary 1 is always satisfied.

Corollary 2 of Hall's Theorem *In an m-by-n matrix of open and closed cells, suppose that each row contains the same number of open cells and each column contains the same number of open cells. Assume also that $n \geq m$. Then the matrix contains m independent open cells.*

M35 Give an example that shows that the condition $n \geq m$ cannot be omitted.

M36 Which of the examples in problem M34 (if any) satisfy the conditions of corollary 2?

M37 Prove corollary 2 by the following steps:

(a) Let h be the number of open cells in each column. Then what is the total number of open cells in the matrix?

(b) Show that each row contains at least h open cells.

More Problems

M38 In this graph, let **M** consist of the three vertical edges.

(a) Construct the corresponding colored digraph and find a green-to-green path.

(b) Use the matching extension algorithm to increase **M** to a larger matching.

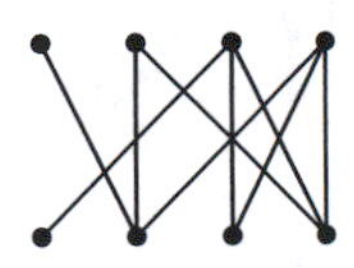

M39 In the graph of problem M7 (shown below), let **M** be the matching consisting of the four heavy edges.

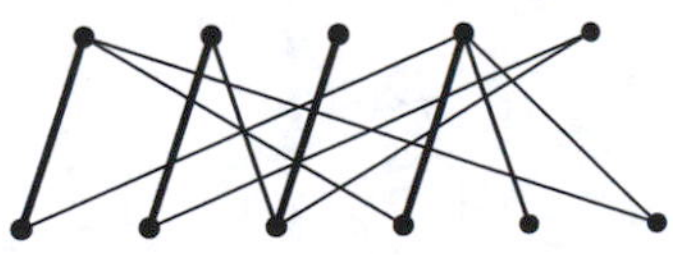

(a) Show that **M** is maximal by constructing the corresponding colored digraph. (Arrows are not necessary. Just remember that edges in **M** go upward, all others downward.)

(b) Find a maximal independent vertex set.

(c) Find a minimal covering vertex set.

M40 In the graph of problem M1 (shown below), let **M** be the matching consisting of the two horizontal edges.

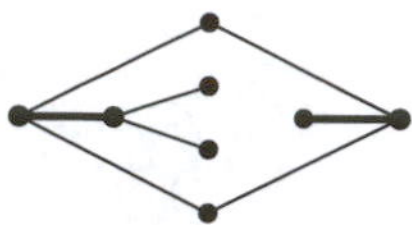

(a) Without re-drawing the graph in two rows, construct the colored digraph, starting with the assumption that the vertex at the left is blue.

(b) Use the matching extension algorithm to increase **M** to a larger matching.

M41 In the graph of problem M4 (shown below), let **M** be the matching consisting of the four heavy edges.

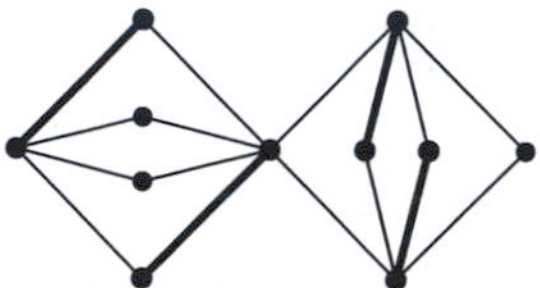

(a) Without re-drawing the graph in two rows, construct the colored digraph, starting with the assumption that the vertex at the left is blue.

(b) Check that there is no green-to-green path. What does that indicate about the matching?

(c) Find the corresponding independent vertex set **S**.

M42 In this graph, let **M** be the matching consisting of the seven heavy edges.

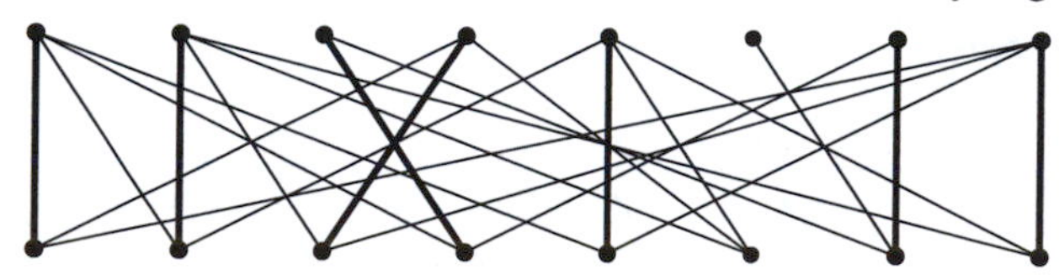

(a) Show that **M** is maximal by constructing the corresponding colored digraph. (Arrows are not necessary. Just remember that edges in **M** go upward, all others downward.)

(b) Find a maximal independent vertex set.

(c) Find a minimal covering vertex set.

M43 Starting with the matching consisting of the five vertical edges in this graph,

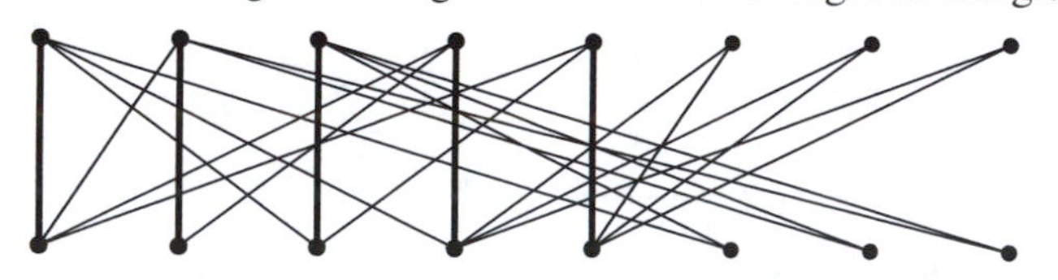

(a) Use the colored digraph construction as many times as necessary to produce a maximal matching.

(b) Find a maximal independent vertex set.

(c) Find a minimal covering vertex set.

M44 State and prove a max/min principle for independent vertex sets and covering edge sets in any graph.

M45 Find five independent open cells in this matrix. Does either corollary of Hall's Theorem apply?

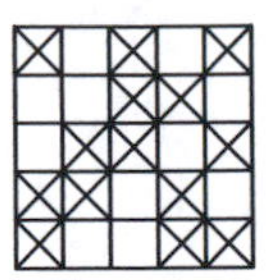

M46 (a) Give a reason why this matrix must contain eight independent open cells.

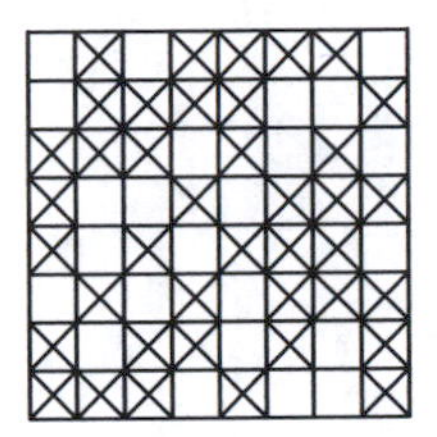

(b) Now find them.

M47 (a) What does a covering edge set correspond to in a matrix of open and closed cells?

(b) Answer the same question for an independent vertex set.

M48 Translate the form of the König–Egervary Theorem that says $\alpha = \beta'$ into matrix terms.

M49 Translate Hall's Theorem and its two corollaries into bipartite graph terms.

M50 (a) In an m-by-n matrix of open and closed cells, suppose that for each integer k, $2 \leq k \leq m$, every combination of k rows contains open cells in at least $k - 1$ different columns. Prove that the matrix contains at least $m - 1$ independent open cells. (Hint: Add a column.)

(b) Generalize (a) as follows: For a fixed integer r, $1 \leq r \leq m$, suppose that for each integer k, $r \leq k \leq m$, every combination of k rows contains open cells in at least $k - r$ different columns. Then what can you conclude? How can you prove this?

Applications of Matching Theory

In this chapter we look at a variety of applications of Hall's Theorem and the König-Egervary Theorem.

Sets and Representatives

Below are six 3-element sets, represented as 3-letter words. The problem is to select one letter from each set in such a way that no letter is selected more than once.

$$BDE \quad CDE \quad ACE \quad BCF \quad ABF \quad ADF$$

N1 Do this by trial and error.

N2 Find an interpretation of this problem in terms of a matrix of open and closed cells. Here is a generous hint: ⊠□⊠□□⊠

N3 Restate the same problem in terms of a bipartite graph.

Interpreting the sets as committees formed from a group of people (the letters), we can think of the selected letters as "representatives" from the committees. In these terms, problem N1 asks for a choice of distinct representatives including one from each committee.

N4 Apply a corollary of Hall's Theorem to the situation of problem N1 to predict that a solution exists.

Definition A *system of distinct representatives (SDR)* for a given set of committees is a choice of distinct members from distinct committees. An SDR that includes a representative from each committee is called a *complete system of distinct representatives*, or a *complete SDR*.

For example, the underlined letters below form an incomplete SDR for the committees in problem N1.

$$\underline{B}DE \quad \underline{C}DE \quad \underline{A}CE \quad BC\underline{F} \quad ABF \quad A\underline{D}F$$

In general, a system of distinct representatives, not necessarily complete, corresponds to a set of independent open cells in a matrix or equivalently, to a matching in a bipartite graph. In the latter interpretation, each edge in the graph represents the membership of one person on one committee, and the edges in a matching represent a choice of representatives.

N5 Try to find a complete SDR for these committees. When you decide that it can't be done, explain why.

$$EH \quad BD \quad CH \quad AE \quad FG \quad CE \quad AH$$

N6 Find a largest possible SDR for the committees in problem N5 and use the result to determine a maximal set of independent open cells in the corresponding matrix.

N7 (a) Restate Hall's Theorem for a set of m committees formed from a group of n people, to give a necessary and sufficient condition for the existence of a complete SDR. Use the word "union".

(b) Indicate how this applies to problem N5.

N8 Translate the two corollaries of Hall's Theorem into committee terms.

Latin Squares

A *latin square* is a square matrix in which each element from a given set appears exactly once in every row and column.

Example

A	B	C	D
C	A	D	B
B	D	A	C
D	C	B	A

A *partial latin square* is any matrix, not necessarily square, in which no element appears more than once in any row or column.

N9 Add two columns to this partial latin square so that it becomes a 5-by-5 latin square.

A	B	C		
B	D	E		
C	A	B		
D	E	A		
E	C	D		

N10 Select a complete system of distinct representatives from the five committees shown below. What does this have to do with the preceding problem?

$$DE \quad AC \quad DE \quad BC \quad AB$$

N11 In general, suppose you want to extend a partial latin square to include one more column. Indicate how you might try to do this by looking at an appropriate set of committees.

N12 Suppose that an m-by-n partial latin square, with $m > n$, involves exactly m different elements. Show that the matrix can be extended to include one more column without introducing any new elements, by the following steps:

(a) Define the appropriate committees. How many are there? How many members are on each committee?

(b) Show that each person is on the same number of committees.

(c) It follows that a complete SDR exists. Why?

N13 Problem **N12** implies that an m-by-n partial latin square involving m different elements, where $m > n$, can always be extended to an m-by-m latin square. Explain how it shows this.

Permutation Matrices

A *permutation matrix* is a square matrix in which each line (row or column) contains exactly one 1 and all other entries in the line are 0.

Example

0	1	0	0
0	0	0	1
0	0	1	0
1	0	0	0

It is easy to see that the sum of any number of permutation matrices is a matrix in which each line sum is the same. However, what about the converse of this statement? Specifically, we have this question:

Suppose that each line in a square matrix of nonnegative integers has the same sum. Then is the matrix necessarily a sum of permutation matrices?

N14 In this problem we will write the matrix below as a sum of four permutation matrices.

1	1	1	1
0	2	0	2
2	0	1	1
1	1	2	0

(a) What is the significance of this matrix of open and closed cells? Find four independent open cells.

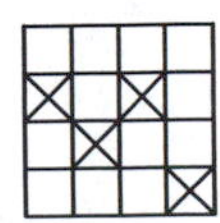

Use these cells to determine a permutation matrix P_1.

(b) Subtract P_1 from the given matrix and repeat the process. Continue until you obtain the desired result.

From this example, it is clear how the process would work in general, as long as enough independent open cells exist on each step. An application of Hall's Theorem will show that they exist.

Let M be an n-by-n matrix of nonnegative integers in which each line sum is equal to some positive integer r. Form a matrix N of open and closed cells in which open cells correspond to positive entries in M, closed cells to zeros. We claim that N contains n independent open cells.

Applying Hall's Theorem, we have to show that for each k, $1 \leq k \leq n$, every combination of k rows of N contains open cells in at least k different columns. For convenience and without loss of generality, we assume that the k rows being considered are at the top of the matrix and that in these rows, all open cells occur in the first h columns.

Let M_1 be the k-by-h matrix consisting of the first k rows and the first h columns of M.

N15 (a) What is the sum of the elements in any row of M_1?

(b) What can you say about the sum of the elements in any column of M_1?

(c) It follows that $h \geq k$. Why?

(d) What does this imply about N?

(e) What does it tell you about M?

We conclude that the process used in problem N14 will always work for a matrix of nonnegative integers in which each line sum is the same, resulting in a sum of permutation matrices.

Problems N35–36 show that a similar result applies if the integers are replaced by nonnegative real numbers.

N16 Restate what has been proved in terms of edge coloring in a regular bipartite multigraph. Do we already know this? (Refer back to problem L32 if necessary.) How does it apply to this example?

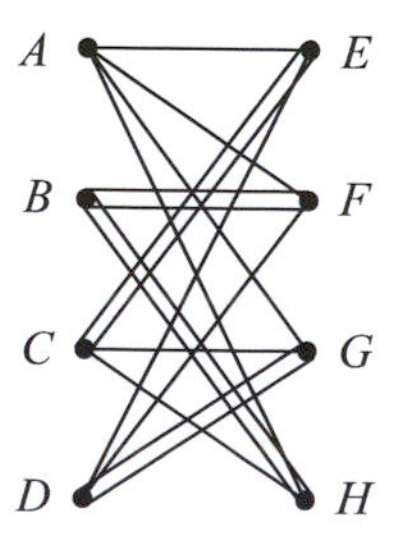

The Optimal Assignment Problem

Suppose that we have to assign jobs to a group of workers, and each worker is rated on a scale of zero to 10 according to that worker's competence for each job, with 10 being highest. Below is an example.

	Bill	Ross	George	Al	Ralph
used car salesman	8	10	4	5	0
fraternity president	9	2	8	4	3
standup comic	7	4	1	5	9
talk show host	8	1	3	2	5

Since there are five workers for only four jobs, one of the workers will remain unemployed. So that we can work with a square matrix, we add a fifth row in which each worker is given an equal rating.

unemployed	10	10	10	10	10

Each job is to be assigned to a different worker and we want the total competence in the assignment to be maximized. For example, the assignment along the main diagonal would result in a total competence of 23. Maybe there is a better choice. We will show how an optimal assignment (one that maximizes total competence) can be constructed by a process known as the *Hungarian Algorithm.*

The problem is usually stated in terms of minimizing a quantity, so we'll convert the competence ratings to costs by subtracting each rating from 10, and then try to minimize the total cost. It is clear that an assignment that minimizes the total cost will also maximize the total competence.

The resulting cost matrix is shown below.

$$\begin{array}{|ccccc|} 2 & 0 & 6 & 5 & 10 \\ 1 & 8 & 2 & 6 & 7 \\ 3 & 6 & 9 & 5 & 1 \\ 2 & 9 & 7 & 8 & 5 \\ 0 & 0 & 0 & 0 & 0 \end{array}$$

In the next step we locate the minimum element in each row and subtract this value from all entries in that row.

$$\begin{array}{|ccccc|} 2 & 0 & 6 & 5 & 10 \\ 0 & 7 & 1 & 5 & 6 \\ 2 & 5 & 8 & 4 & 0 \\ 0 & 7 & 5 & 6 & 3 \\ 0 & 0 & 0 & 0 & 0 \end{array}$$

N17 Explain why this reduction does not change the optimal assignment. In general, show that adding a constant to all entries in any line of a square cost matrix does not affect which assignment is optimal.

N18 Using the last matrix above, find an optimal assignment. Who is assigned to each job?

In the process shown above, the cost matrix was reduced in such a way that each line contained a zero as its minimum element, and then an optimal assignment became reasonably easy to find by inspection. For more complicated situations, however, we need a systematic method for arriving at an optimal assignment. That method is the Hungarian Algorithm, which we illustrate by continuing with the same example.

The last cost matrix above does not contain five independent zeros. It follows that all of the zeros in this matrix can be covered by fewer than five lines.

N19 (a) How does this follow? What theorem are we using here?

(b) Find such a set of lines.

The line set **L** found above divides the cells in the cost matrix into three groups: uncovered cells, twice-covered cells, and once-covered cells. Let m denote the minimum of all costs in uncovered cells and do this:

the Hungarian step:

Subtract m from all costs in uncovered cells. Add m to all costs in twice-covered cells.

N20 See what happens. Find five independent zeros in the new matrix. Is the optimal assignment still the same?

It is not immediately obvious that changing the cost matrix by the Hungarian step will always leave the optimal assignment unchanged, but it is true. The next problem explains why.

N21 Show that the Hungarian step is equivalent to this:

Add m to all costs in rows of **L**; Add m to all costs in columns of **L**; Subtract m from all costs in the matrix.

How does that prove that the optimal assignment remains unchanged?

The Hungarian Algorithm consists of repeated application of this step in order to increase the number of independent zeros in a given n-by-n matrix of nonnegative costs. The goal is to arrive at a nonnegative matrix that contains n independent zeros, at which point the process terminates.

N22 What does that accomplish?

N23 Apply the Hungarian Algorithm to this matrix. How many steps does it take?

$$\begin{array}{|cccc|}\hline 0 & 0 & 0 & 0 \\ 0 & 1 & 2 & 2 \\ 0 & 2 & 2 & 2 \\ 0 & 2 & 2 & 2 \\ \hline\end{array}$$

Returning to the general case of an n-by-n cost matrix, we still need to answer one important question: How do we know that n independent zeros will eventually appear in the matrix? The preceding problem suggests that the maximum number of independent zeros increases on each step, but problem N39 will show that this is not always the case. Instead, we take an indirect approach.

We claim that the Hungarian Algorithm always terminates in a finite number of steps. The key to proving this is to show that the sum of all costs in the matrix strictly decreases on each step.

N24 How will that imply that the algorithm eventually terminates? Remember that all costs are assumed to be nonnegative integers.

N25 What does this have to do with the existence of n independent zeros?

The remaining details needed to complete the proof are contained in the next problem. Starting with an n-by-n cost matrix that does not contain n independent zeros, let $\mathbf{L}$ be a line set that covers all of the zeros and contains fewer than n lines. The König-Egervary Theorem shows that such a set exists. Let r denote the number of rows in $\mathbf{L}$ and c the number of columns. Then $r + c < n$.

N26 (a) In terms of r and c, how many twice-covered costs does the matrix contain?

(b) How many uncovered costs are there?

(c) Compare your answers to (a) and (b). Which one is always larger? Prove it.

(d) What does the Hungarian step do to the sum of all costs in the matrix?

That completes the proof that the Hungarian Algorithm must eventually terminate, producing n independent zeros, for any n-by-n matrix of nonnegative integer costs. The cells that contain these zeros represent an optimal assignment.

More Problems

N27 Find a maximal (not necessarily complete) SDR for each set of committees below.

(a)	(b)
ABF	BEH
DEG	CFG
ADF	BDE
CEH	AEH
DFG	ACG
BCH	BDH
BGH	ABE
ACE	ADH

N28 For each set of committees below, decide whether one or both of the corollaries of Hall's Theorem applies. In each case, find a complete SDR if it exists.

(a)	(b)	(c)
CDF	$ACDG$	$ABEGH$
ABE	$ADEH$	$CDEFH$
BEF	$BCFH$	$BDEGH$
ABD	$ACGH$	$ABCFG$
CEF	$BDEF$	$ACDEF$
ACE	$BEFG$	$ADEGH$

N29 The König-Egervary Theorem, when applied to a set of committees, says that the maximum number of distinct representatives is equal to ...what? The difficulty is that the concept of a covering line set does not translate easily to the committee situation. Try to put it into words.

N30 (a) Find a 3-by-3 partial latin square involving four symbols, that cannot be extended to a 4-by-4 latin square.

(b) Extend it to a 6-by-6 latin square.

N31 Show that it is impossible to extend this matrix to a 7-by-7 partial latin square without introducing any new symbols.

C	B	D	G	F	A
H	E	G	F	A	C
G	F	A	E	B	D
D	G	F	H	C	B
B	D	C	A	E	H
F	A	H	B	D	G
E	C	B	D	G	F

N32 (a) Extend this matrix to a 4-by-7 partial latin square without introducing any new symbols. However, show that the SDR method of problem N11 could result in a choice of a fifth column that makes it impossible to complete the matrix.

A	B	C	D			
B	C	D	E			
D	A	B	G			
C	F	E	A			

(b) Color the open cells in this matrix using three colors, with no color appearing twice in any line. Indicate what this has to do with König's Coloring Theorem (Edge Coloring Theorem #3 in chapter L).

A	B	C	D	E	F	G
X	X	X	X			
	X	X	X	X		
X	X		X			X
X		X		X	X	

c) How can the result of (b) be used to complete the matrix in (a)?

N33 (a) Use the method of the preceding problem to extend the matrix in problem N31 to a 7-by-13 partial latin square containing 13 different symbols.

(b) The same matrix can then be extended to a 13-by-13 latin square. How would you know this?

N34 Generalize the result of the preceding problem to show that every m-by-n partial latin square can be extended to an $(m+n)$-by-$(m+n)$ latin square.

N35 Write this matrix as a linear combination of permutation matrices with positive real coefficients.

$$\begin{array}{cccc} \pi & 0 & e & 0 \\ 0 & \pi & e & 0 \\ e & \sqrt{7} & 0 & \pi - \sqrt{7} \\ 0 & e - \sqrt{7} & \pi - e & e + \sqrt{7} \end{array}$$

N36 State and prove a theorem about square matrices with nonnegative real entries and equal line sums.

N37 Apply the Hungarian Algorithm to this cost matrix to find an optimal assignment. First transform the matrix so that each row and column contains at least one zero.

$$\begin{array}{ccccc} 2 & 6 & 4 & 0 & 3 \\ 1 & 5 & 2 & 4 & 6 \\ 0 & 2 & 5 & 1 & 2 \\ 4 & 1 & 3 & 2 & 5 \\ 6 & 2 & 4 & 1 & 7 \end{array}$$

N38 Suppose that a square cost matrix contains one or more negative entries. Figure out how to deal with this situation.

N39 Apply one step of the Hungarian Algorithm to this cost matrix and see what happens to the maximum number of independent zeros.

$$\begin{array}{cccc} 0 & 0 & 0 & 0 \\ 0 & 0 & 1 & 2 \\ 0 & 1 & 2 & 2 \\ 0 & 2 & 2 & 2 \end{array}$$

N40 The covering line set used in a step of the Hungarian Algorithm is not required to be minimal. Give a practical reason why we might not want to require that.

N41 In this problem we will show that the maximum number of independent zeros in a cost matrix does not decrease on a step of the Hungarian Algorithm that uses a minimal covering line set. Let **L** be a minimal line set that covers all zeros in the matrix and let **Z** be a maximal set of independent zeros.

(a) What does the König-Egervary Theorem tell you about the number of elements in each of these sets?

(b) Show that no zero in **Z** is covered by more than one line in **L**. (Hint: There would not be enough lines in L to cover all of the remaining members of **Z**.)

(c) All zeros in **Z** remain in the matrix after the step is completed. Why?

Although an optimal assignment of n jobs to n workers does not necessarily assign every job to a worker who is most qualified for that job, the next two problems will show that there is always at least one job that goes to the most qualified worker. To put this another way, any assignment that assigns no job to the most qualified worker cannot be optimal. We will prove this by showing how such an assignment can be improved.

N42 In the matrix below, the circled costs represent the present assignment of jobs to workers. Notice that no job is assigned to the most qualified (minimum cost) worker.

	W_1	W_2	W_3	W_4
J_1	③	5	2	4
J_2	4	②	0	6
J_3	5	6	④	1
J_4	0	3	7	②

Obtain a lower cost assignment by the following steps:

(a) Construct a bipartite graph in which each job vertex J_i has degree 2, with one edge going to the presently assigned worker W_i and the other going to a more qualified worker W_j.

J_1 • • W_1

J_2 • • W_2

J_3 • • W_3

J_4 • • W_4

(b) Find a cycle in the resulting graph and figure out how to use it to improve the assignment.

N43 Use the method of the preceding problem to prove that in the general case, any assignment that assigns no job to the most qualified worker can be improved. How do you know that the graph contains a cycle? (Hint: Consider the number of edges and vertices in the graph and refer back to chapter D if necessary.)

N44 In an optimal assignment of n jobs to n workers, must there always be at least one worker who is assigned to his/her best job?

O

Cycle-Free Digraphs

A *cycle-free digraph* is a directed graph in which there are no directed cycles. For example, the first digraph below is cycle-free. The second one is not.

Notation For vertices A and B in a cycle-free digraph $\mathbf{D}$, we write $A \to B$ to indicate that a directed path from A to B exists in $\mathbf{D}$. In other words, B is reachable from A.

Chains and Antichains

A *chain* in a cycle-free digraph is a sequence of vertices with the property that each vertex in the sequence is reachable from the preceding one. The *length* of a chain is the number of vertices in the chain. For example, a chain of length 1 is a single vertex. A chain of length 3 consists of vertices A, B, and C such that $A \to B$ and $B \to C$. To put it another way, the vertices of a chain occur along some directed path, although there may be other vertices between them on the path.

O1 Which of these are chains in the digraph below?

$ABCD \qquad AD \qquad EFGH \qquad EH$

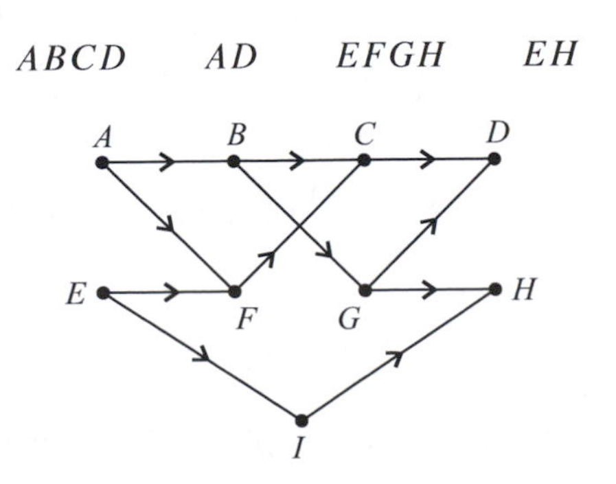

An *antichain* is a set of mutually unreachable vertices: For any A and B in an antichain, $A \nrightarrow B$ and $B \nrightarrow A$. The *order* of an antichain is the number of vertices in the antichain.

O2 Find an antichain of order 3 in the digraph of problem O1.

Some interesting digraphs are defined in terms of other mathematical structures.

O3 Vertices in the digraph below are points in the xy-plane with integer coordinates from 0 to 4. Edges go one unit up and one unit to the right.

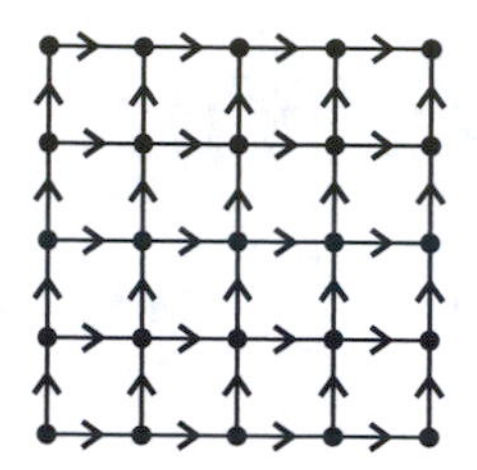

Find a maximal chain and a maximal antichain in this digraph.

O4 Let **D** be a digraph in which vertices are the 16 subsets of $\{1, 2, 3, 4\}$. Edges from a set A go to each set B that can be obtained from A by adding one more element.

(a) What does $A \rightarrow B$ indicate in this case?

(b) Find a chain of length 5 and an antichain of order 6.

O5 Let **D** be a digraph in which vertices are the positive integers from 1 to 100. Edges from an integer n go to each integer ≤ 100 that can be obtained by multiplying n by a prime.

(a) What does $A \rightarrow B$ indicate in this case?

(b) Find a chain of length 7.

(c) There is an antichain of order 50. Can you find it?

Chain Decompositions

A *chain decomposition* of a cycle-free digraph is a partition of the vertices of **D** into disjoint chains. The *order* of a chain decomposition is the number of chains.

For example, one chain decomposition of the digraph in problem O1 consists of the four chains $ABCD$, EF, GH and I.

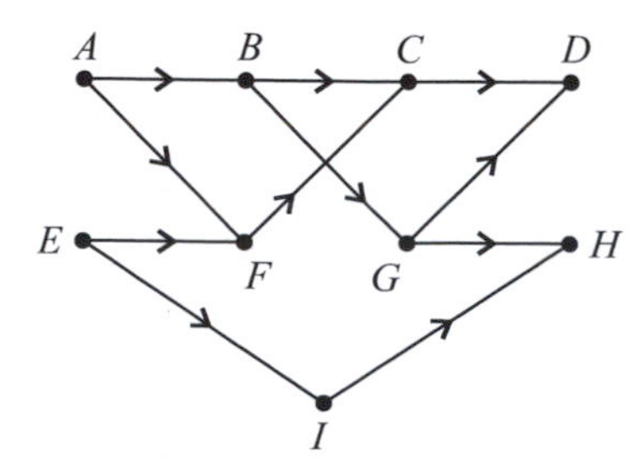

O6 Find a chain decomposition of order 3 in the same digraph.

O7 Use the fact that F, G, and I are mutually unreachable vertices to prove that the decomposition found in problem O6 is minimal: that is, **D** has no chain decomposition of order less than 3.

O8 Complete this statement and explain why it is true: If a cycle-free digraph contains an antichain of order k, then any chain decomposition of the digraph must contain at least ________________.

O9 State and prove a max/min principle for antichains and chain decompositions in a cycle-free digraph, similar to the ones for bipartite graphs in chapter M.

O10 Find a minimal chain decomposition for the digraph in problem O3.

O11 Find a maximal antichain and a minimal chain decomposition for this digraph.

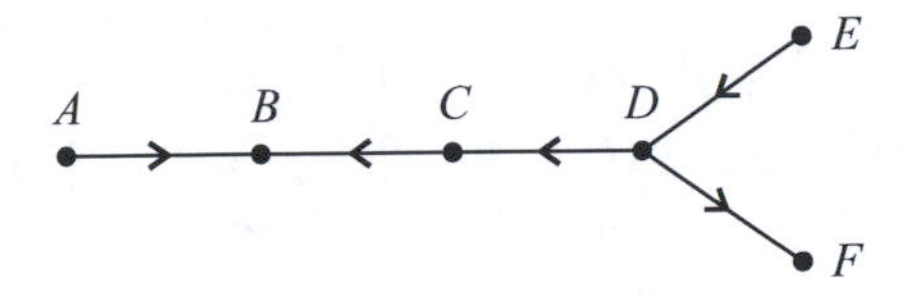

O12 Without looking at the theorem below, guess what Dilworth proved.

Dilworth's Theorem *In any cycle-free digraph, there exists an antichain and a chain decomposition of equal order. The antichain is maximal and the chain decomposition is minimal.*

O13 Find a maximal antichain and a minimal chain decomposition for the digraph in problem O4. (Suggestion: Use abbreviated set notation, such as 123 for the set $\{1, 2, 3\}$.)

O14 Let **D** be a digraph in which vertices are positive integers, as in problem O5, except restricted to only 1 through 12. Find a maximal antichain and a minimal chain decomposition for **D**.

O15 Find a maximal antichain and a minimal chain decomposition for this digraph.

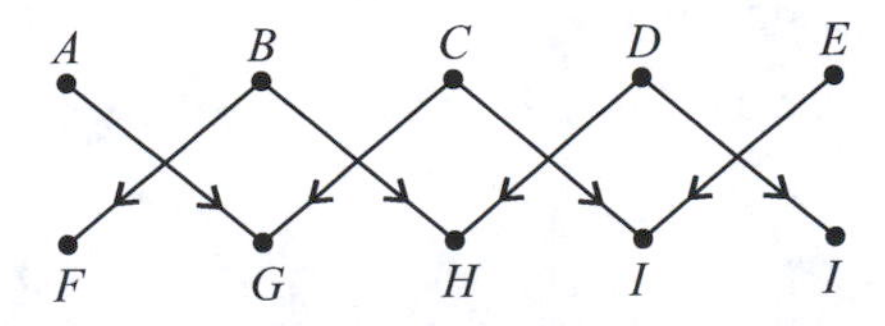

Not surprisingly, there is a connection between Dilworth's Theorem and the König–Egervary Theorem, which (in the form $\alpha' = \beta$ or $\alpha = \beta'$) also says that the maximum of one quantity is equal to the minimum of another quantity. However the connection is not a very obvious one. For example, the members of an antichain do not correspond directly to the members of a matching. We will prove Dilworth's Theorem by applying the König–Egervary Theorem to an appropriate bipartite graph. Before getting into this, however, we mention one consequence of Dilworth's Theorem.

Corollary *Let m_1 be the order of a maximal antichain and let m_2 be the length of a maximal chain in a cycle-free digraph containing n vertices. Then the product $m_1 m_2$ is greater than or equal to n.*

O16 Use Dilworth's Theorem to prove this.

Proof of Dilworth's Theorem

The first step is to construct a bipartite graph **B** in which there are two vertices X_1 and X_2 for each vertex X in **D**. An edge joins vertices X_1 and Y_2 whenever X and Y are two distinct vertices in **D** such that $X \to Y$.

For example, since $A \to B$ in this digraph **D**, A_1 and B_2 are adjacent in **B**.

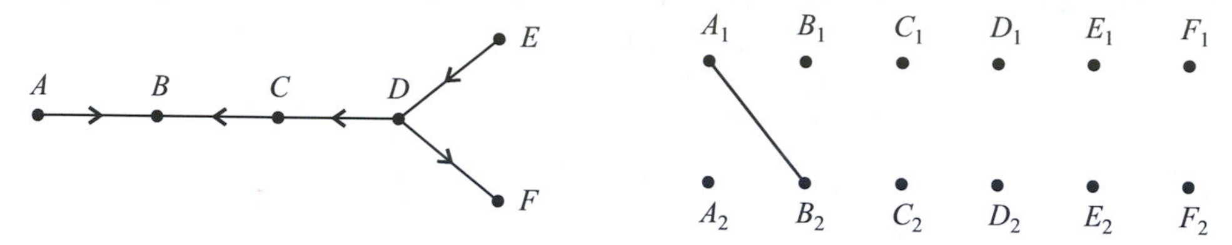

O17 (a) Fill in the remaining edges of **B** and find a minimal covering vertex set **V**. Use a max/min principle to verify that **V** is minimal.

(b) Find all vertices X in **D** such that X_1 is not in **V** and X_2 is not in **V**. What do you notice about this set?

Returning to the general case of a cycle-free digraph **D** and the corresponding bipartite graph **B**, let **V** be a minimal covering vertex set in **B**.

O18 Show that if X and Y are two vertices of **D** such that **V** contains none of the vertices X_1, X_2, Y_1 and Y_2, then X and Y are mutually unreachable in **D**.

O19 Prove that **D** contains an antichain consisting of at least $n - \beta$ vertices, where n is the number of vertices in **D** and β is the vertex covering number for **B**.

Problem O19 represents one half of what we need to prove Dilworth's Theorem. The other half comes from looking at a maximal matching in **B**.

O20 Returning to the example in problem O17,

(a) Find a maximal matching in **B** and add those edges to the graph below.

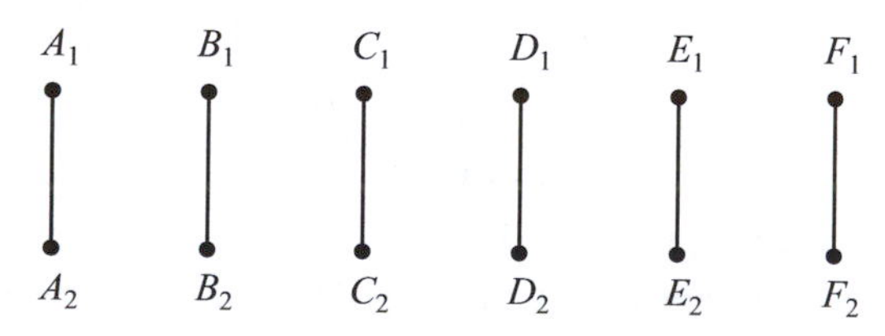

Use the resulting graph to find a chain decomposition of **D**.

(b) Now start over with a different maximal matching and find another chain decomposition.

Problem O20 illustrates the process by which a chain decomposition of a cycle-free digraph **D** can be constructed from a maximal matching in the corresponding bipartite graph **B**. In general, the matching is added to a graph that consists of edges X_1X_2 for each vertex X in **D**. The result is another bipartite graph $\mathbf{B}^*$ in which the degree of each vertex is either 1 or 2.

O21 (a) Why is the degree of each vertex in $\mathbf{B}^*$ either 1 or 2?

(b) Explain why each component of $\mathbf{B}^*$ must be a simple path. In particular, why can't $\mathbf{B}^*$ contain a cycle?

O22 (a) Let n be the number of vertices in **D** and let α' be the edge independence number for **B**. Show that the number of vertices of degree 1 in $\mathbf{B}^*$ is $2n - 2\alpha'$.

(b) How many components does $\mathbf{B}^*$ contain?

We need to know one more thing before we can say that the components of $\mathbf{B}^*$ correspond to chains in a chain decomposition of **D**:

O23 In each component of $\mathbf{B}^*$ the edges alternate between vertical edges X_1X_2 and non-vertical edges X_1Y_2, where $X \neq Y$. Explain why this is true. Which type of edge occurs at each end of the path?

O24 Now convince yourself that the components of $\mathbf{B}^*$ correspond to chains in **D** and that the result is a chain decomposition.

Summarizing, we have established this:

If **D** is a cycle-free digraph containing n vertices, then **D** contains an antichain consisting of at least $n - \beta$ vertices (problem O19) and a chain decomposition consisting of $n - \alpha'$ chains (problems O22–24), where β is the vertex covering number and α' is the edge independence number of the bipartite graph **B**.

O25 Use the above to complete the proof of Dilworth's Theorem. Do we know that the antichain is maximal? Can it contain more than $n - \beta$ vertices?

O26 Let **D** be the digraph below. Fill in the edges of the corresponding bipartite graph **B** and construct $\mathbf{B}^*$ using an appropriate matching. Use these graphs to obtain a maximal antichain and a minimal chain decomposition for **D**.

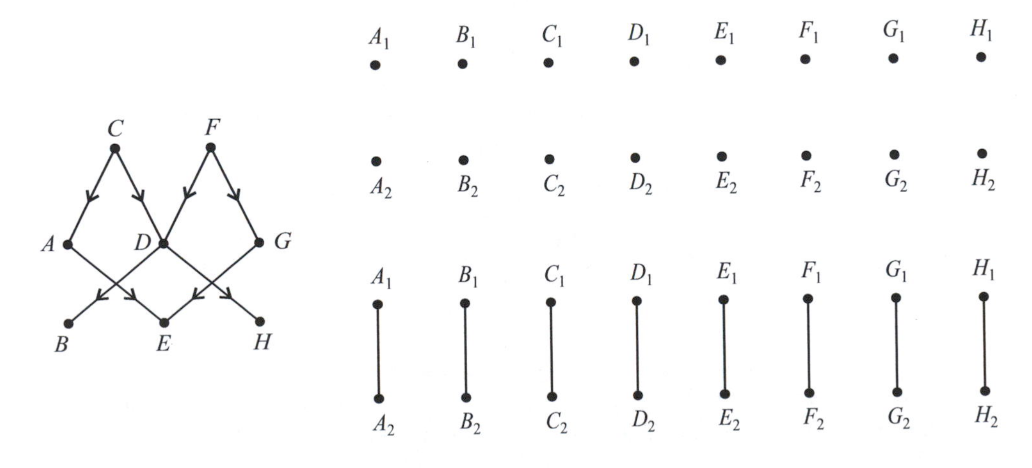

More Problems

O27 In the digraph of problem **O1**, can $ABCD$ be one of the chains in a minimal chain decomposition?

O28 The set $\{1, 2, 3, 4, 5\}$ contains ten 2-element subsets and ten 3-element subsets. Find a one-to-one correspondence between these two collections of subsets in which each 2-element set corresponds to a 3-element set that contains it. But first use a theorem on bipartite graphs to predict that such a correspondence exists. The diagram below is a hint.

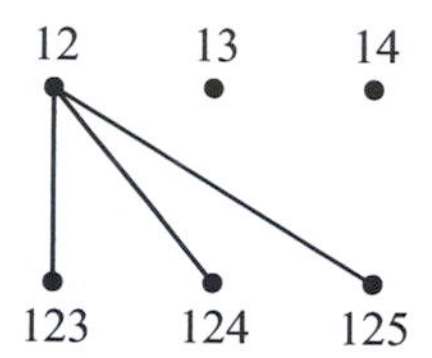

O29 Let **D** be a digraph in which vertices are the 32 subsets of $\{1, 2, 3, 4, 5\}$. Edges from a set A go to each set B that can be obtained from A by adding one more element, as in problem **O4**. Find a maximal antichain and a minimal chain decomposition for **D**. The preceding problem may help.

O30 Find a chain decomposition of order 50 in the digraph of problem **O5** in which vertices are the positive integers from 1 to 100. Do we know that the decomposition is minimal? (Hint: Put each odd number in a different chain.)

O31 Let **D** be a digraph in which vertices are the 27 points in xyz-space with integer coordinates from 0 to 2. Edges go from (x, y, z) to $(x + 1, y, z)$, $(x, y + 1, z)$, and $(x, y, z + 1)$ as long as coordinates do not exceed 2.

(a) Find an antichain of order 6.

(b) Find a chain decomposition of order 9.

(c) Now try to find a maximal antichain and a minimal chain decomposition.

O32 Let **D** be the digraph from problem **O1**, shown below. Use bipartite graphs **B** and **B*** as in the proof of Dilworth's Theorem to obtain a maximal antichain and a minimal chain decomposition for **D**.

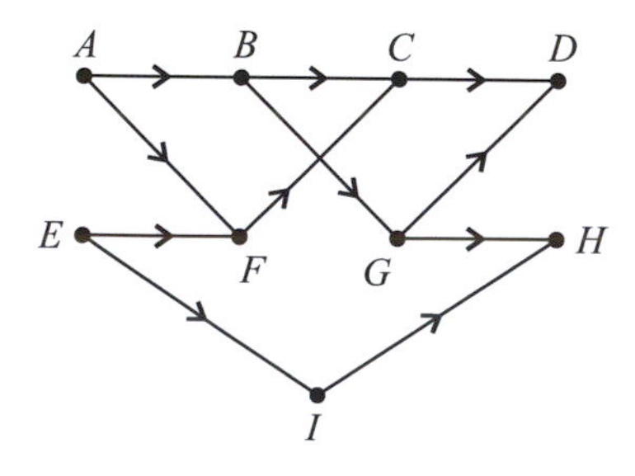

O33 For a given cycle-free digraph **D**, define the *unreachability graph* of **D** to be a graph **G** whose vertices are the same as those of **D**. Vertices A and B are adjacent in **G** if they are mutually unreachable in **D**: $A \nrightarrow B$ and $B \nrightarrow A$.

(a) What does an antichain in **D** correspond to in **G**?

(b) What does a chain correspond to?

(c) Find a relationship between chain decompositions of **D** and proper vertex colorings of **G**.

O34 (a) Recall the definition of the *clique number* γ of a graph. (Refer back to problem I27 if necessary.) How is γ related to the vertex coloring number χ in any graph?

(b) Dilworth's Theorem establishes a stronger relation between γ and χ for the unreachability graph of a cycle-free digraph. What is this relation?

(c) Show that an unreachability graph can never be isomorphic to the cycle graph $\mathbf{C}_5$.

O35 Construct the unreachability graph corresponding to the digraph from problem O17, shown below.

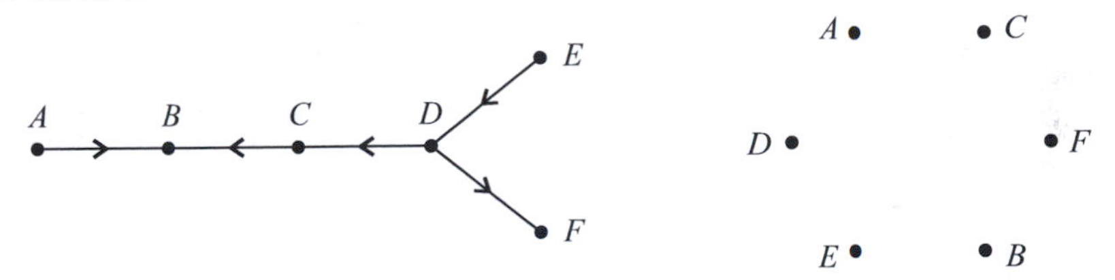

Find a maximal clique and a minimal proper vertex coloring in this graph.

O36 Define the *height* of a vertex in a cycle-free digraph to be the length of a longest chain ending at that vertex.

(a) Show that vertices of equal height are mutually unreachable.

(b) Use the concept of height to give a second proof of the corollary to Dilworth's Theorem. (Hint: How many possible heights are there?)

O37 Let m and n be two positive integers and let **D** be a cycle-free digraph that contains more than mn vertices. Use the corollary to Dilworth's Theorem to prove that **D** contains either a chain of length $m + 1$ or an antichain of order $n + 1$.

O38 This problem appeared on the 1966 William Lowell Putnam Mathematical Competition: Let $a_1 < a_2 < \cdots < a_{mn+1}$ be $mn + 1$ positive integers. Prove that you can select either $m + 1$ of them, no one of which divides any other, or $n + 1$ of them, each dividing the following one.

O39 (a) Find a sequence of 100 integers that does not contain a monotone (increasing or decreasing, not necessarily strictly) subsequence of length 11.

(b) Let $a_1, a_2, \ldots, a_{101}$ be any sequence of 101 real numbers. Prove that the sequence contains a monotone subsequence of length 11. (Hint: Form a digraph in which the vertices are the points $(1, a_1)$, $(2, a_2), \ldots$ in the xy-plane and an edge goes from (h, a_h) to (k, a_k) if $h < k$ and $a_h \leq a_k$.)

P

Network Flow Theory

A *network* is a graph or multigraph in which any or all of the following additional structure can exist:

Directions of some or all of the edges;
Special vertices designated as *sources* or *sinks*;
Numbers associated with some or all of the edges;
Numbers associated with some or all of the vertices.

For our purposes, at least for now, we will assume that we are dealing with a network in which

All edges are directed;
There is at most one edge in each direction between any two vertices;
There is one source and one sink.

Typically, the vertices and edges of a network are referred to as *nodes* and *arcs,* respectively.

Flows in a Network

A *flow* in a given network **N** consists of numbers associated with the arcs of **N**, satisfying certain conditions described below. Think of a flow as indicating the quantity of something that passes through each arc of **N**. Some real-world applications: traffic through streets, fluid through pipes, current through wires, and products through supply routes.

Mathematically, a flow is just a function that assigns a nonnegative real number to each arc of **N**: $\phi(AB)$ represents the quantity of flow through arc AB. In addition to this, a flow has to satisfy the *flow conservation equations*: For each node A other than the source and sink,

$$\sum \phi(X_i A) = \sum \phi(A Y_j)$$

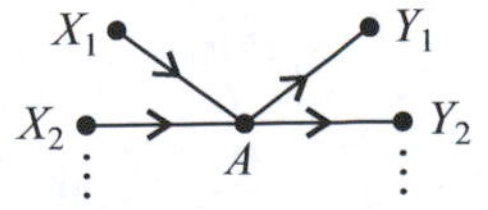

which says that the total inflow to A along incoming arcs is equal to the total outflow along outgoing arcs, or the net inflow to A is zero.

An example of a flow is shown below. In this network, U is the source and V is the sink. The flow conservation equations apply at A, B, and C.

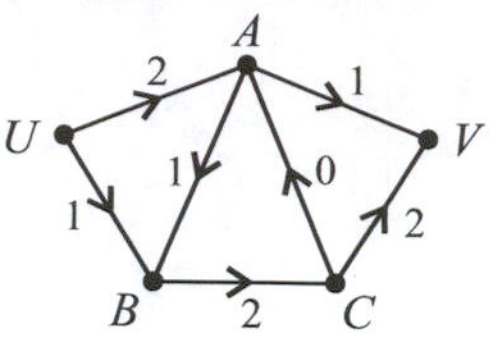

Path flows

A *path flow* is a flow that occurs entirely along a simple path. For example, in the network above, this would be a path flow sending one unit of flow through the path $UBCAV$.

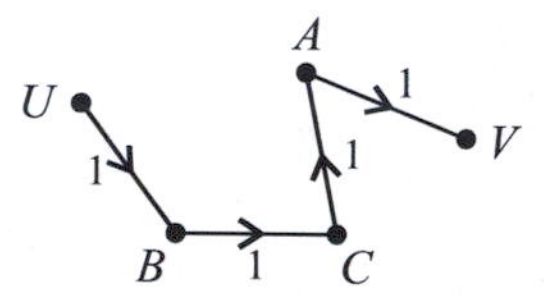

Any quantity of flow can be sent through a path, assigning the same value to each arc of the path and assigning zero to each arc not on the path. We will use the notation $\phi = (UBCAV)$ to indicate the path flow of one unit shown above. Call this a *unit path flow*.

Non-unit path flows can be represented by including a coefficient. For example, a flow of two units through the path $UBCAV$ would be written as $2(UBCAV)$.

P1 Write this flow as a sum of path flows.

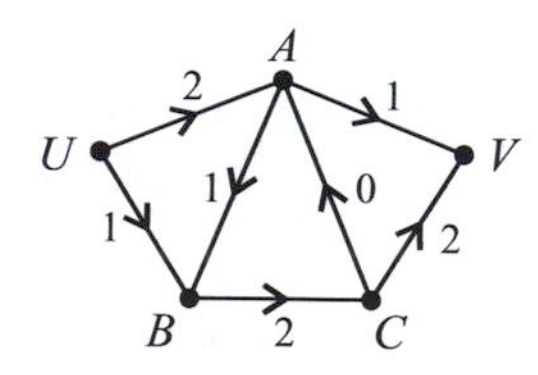

P2 Can this flow be represented as a sum of path flows?

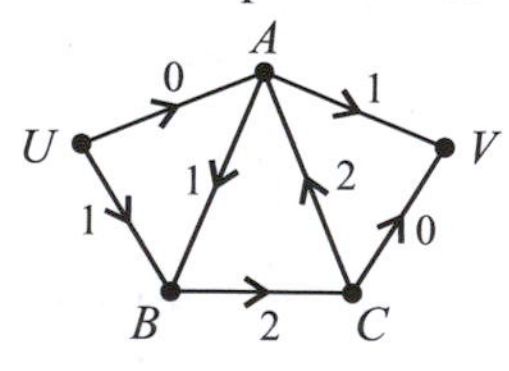

(Note that the endpoints of a path flow have to be the source and the sink.) Guess what is meant by a *cycle flow*.

P3 Represent this flow as a sum of path flows.

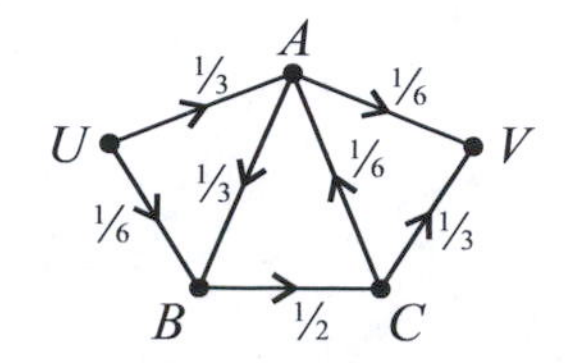

In most cases, we will be dealing with flows in which the quantity of flow through each arc is a nonnegative integer. Such a flow is called an *integral flow*.

The value of a flow

We want to define a quantity associated with a given flow ϕ that in some way indicates how much of whatever the flow represents is going from the source to the sink. We will call this quantity the *value* of ϕ, denoted by the symbol $v(\phi)$. The value of a unit path flow is 1. The value of the flow in problem P1 is 3.

P4 Make up an appropriate general definition for $v(\phi)$ without reference to any physical interpretation of ϕ.

P5 Complete this statement:
Decomposition theorem for integral flows *Let n be a nonnegative integer. Every integral flow of value n can be represented as a sum of n _______ and possibly also some ________.*

P6 What is the value of this flow? Show how the theorem applies in two different ways.

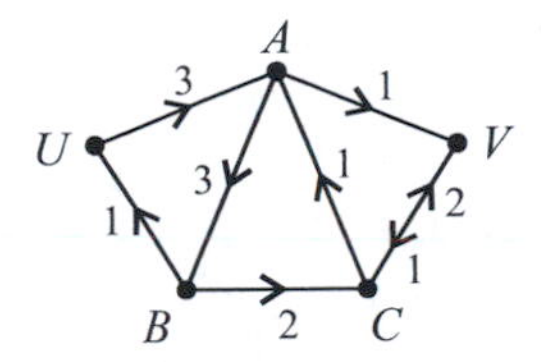

The matrix of a flow

We will represent a flow by an n-by-n matrix, where n is the number of nodes in the network. The example below illustrates how this is done.

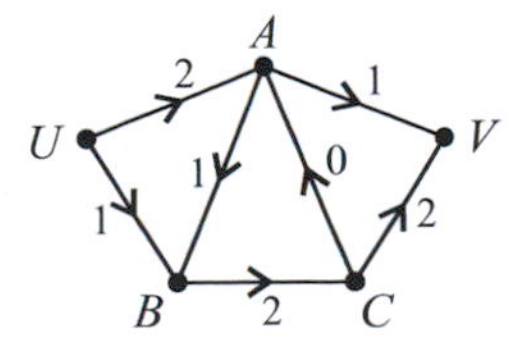

U	0	2	1	0	0
A	−2	0	1	0	1
B	−1	−1	0	2	0
C	0	0	−2	0	2
V	0	−1	0	−2	0

P7 What do the row and column sums represent? Where is the value of the flow represented in the matrix?

Note that the matrix of a flow depends on a particular ordering of the nodes.

To further clarify the definition of the matrix of a flow, we have these examples:

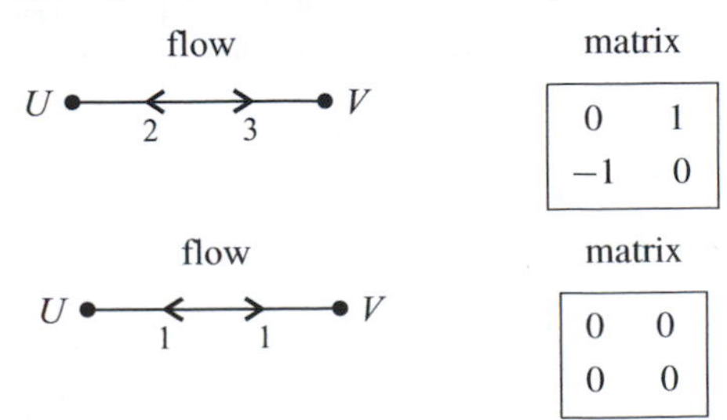

For purposes of the matrix, equal quantities of flow in opposite directions along an arc cancel each other out. This is not the same as saying that the 2-cycle flow (UVU) is the same as the zero flow. But both flows are represented by the zero matrix.

P8 Find the matrix of each flow.

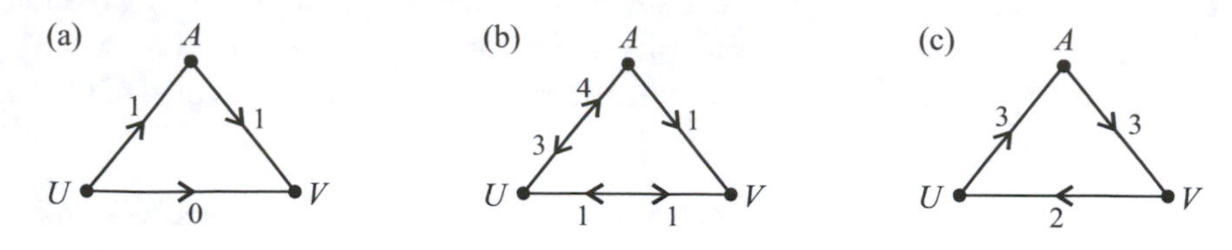

P9 Complete the statement: Two flows correspond to the same matrix if and only if they differ by _______________.

Definition We will call two flows *equivalent* if their matrices are the same. So, for example, any 2-cycle flow is equivalent to the zero flow.

Notation for equivalent flows: $\phi_1 \sim \phi_2$.

Notation for the matrix of a flow: $[\phi]$.

So, in this notation, $\phi_1 \sim \phi_2$ if and only if $[\phi_1] = [\phi_2]$.

Definition A flow is *reduced* if for all nodes A and B, at least one of the quantities $\phi(AB)$ and $\phi(BA)$ is zero (assuming both arcs exist).

P10 Be sure you agree with the following statements:

(1) Every flow is equivalent to a unique reduced flow.

(2) Distinct reduced flows correspond to distinct matrices.

Capacities in a network

Suppose that for each arc AB in a network, there is an upper limit on the quantity of flow that can pass through the arc. This limit, which can be any nonnegative real number, is called the *capacity* of the arc and is denoted by the symbol $c(AB)$.

If we don't want to place any limit on the flow through a particular arc, we can consider its capacity to be infinite.

In physical applications, capacity might represent the number of lanes in one direction along a street, or the cross-section area of a pipe.

Capacities can be represented by placing numbers along arcs or, when the capacities are reasonably small integers, by multiple arrows. For example,

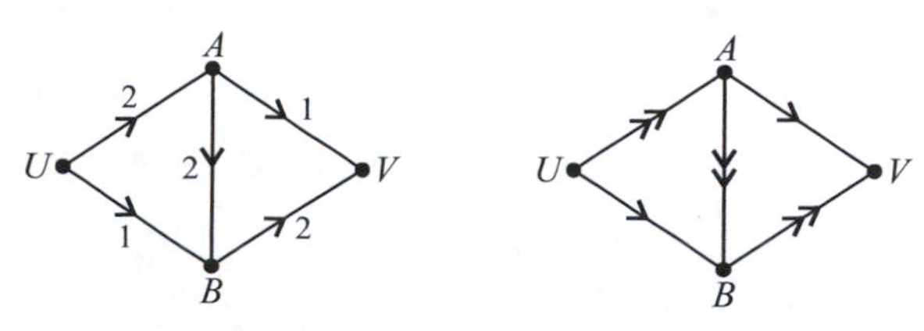

The same information can be represented by the *capacity matrix* for this network, shown below.

	U	A	B	V
U	0	2	1	0
A	0	0	2	1
B	0	0	0	2
V	0	0	0	0

Definition A flow ϕ in a network with capacities is called *feasible* if for each arc AB, $\phi(AB) \leq c(AB)$.

We can now state the main problem involving network flows: in a given network $\mathbf{N}$ with capacities, we want to find a feasible flow ϕ whose value $v(\phi)$ is as large as possible among all feasible flows. Such a flow is called a *maximal flow* in $\mathbf{N}$. Finding a maximal flow in a given network is known as the *maximal flow problem.*

P11 Solve the maximal flow problem for this network. Represent the maximal flow as a sum of path flows.

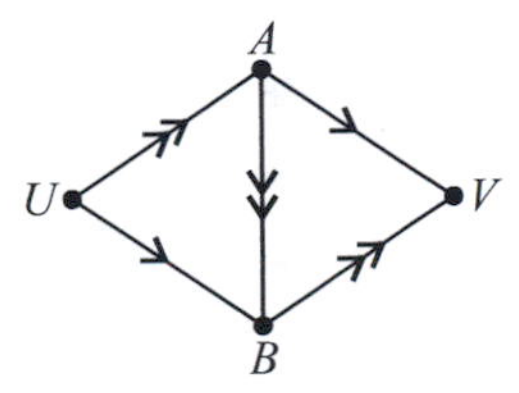

P12 In the same network, find the maximal path flow that can exist in

(a) the path UAV, and

(b) the path $UABV$.

P13 Guess what we would mean by the *capacity of a path*. Guess what a *saturating path flow* probably means.

Maximal flow algorithm (first attempt)

There is a systematic method to solve the maximal flow problem in a given network with capacities. To describe it, we begin with a simple idea that will not usually work. Then we'll see how to modify it to work in all situations.

Starting with a network $\mathbf{N}$, do this:

- Find a path $\mathbf{P}_1$ of positive capacity from the source to the sink;
- Let ϕ_1 be a saturating path flow in $\mathbf{P}_1$;
- Reduce all capacities along $\mathbf{P}_1$ by the value of ϕ_1;
- Call the new network (with reduced capacities) $\mathbf{N}_2$ and repeat the entire process until no path of positive capacity exists from the source to the sink. At that point, add all path flows: $\phi = \phi_1 + \phi_2 + \cdots + \phi_n$.

P14 Use this to do problem P11 again, with $\mathbf{P}_1 = UAV$. Does it work?

P15 Now try the same thing with $\mathbf{P}_1 = UABV$. What goes wrong?

Clearly, it's a mistake to send two units of flow through $UABV$ in this network, but how could we know that from the beginning? Fortunately, we don't have to. We can build into the algorithm a way to go back and change our mind about what gets included in the flow. In other words, the algorithm becomes self-correcting. To do this, we make one crucial change in the algorithm: Instead of removing units of capacity, we *reverse* them. So, returning to problem P15, we get this as network $\mathbf{N}_2$.

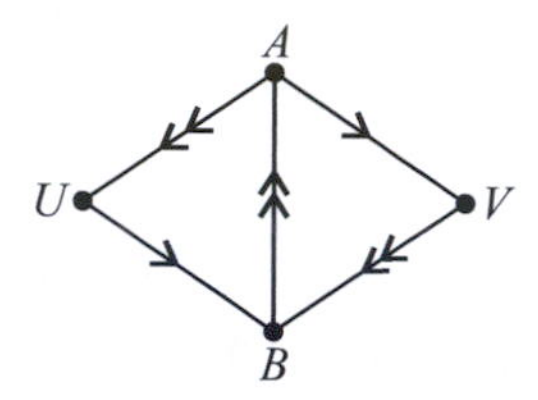

The path $UBAV$ has positive capacity, so we can continue the process by sending one unit of flow through this path: $\phi_2 = (UABV)$. The next network is

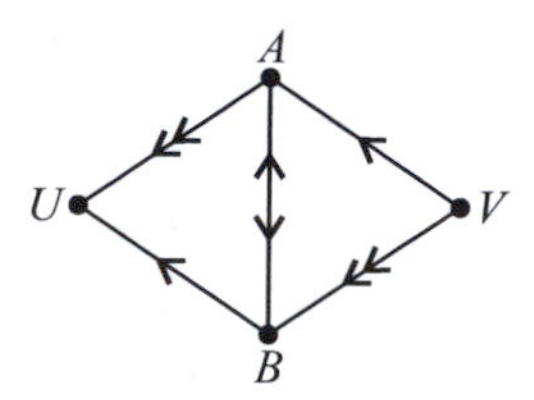

in which no path of positive capacity exists from U to V. The algorithm terminates, producing the flow $\phi = \phi_1 + \phi_2 = 2(UABV) + (UBAV)$.

P16 But that's not a feasible flow in the original network. What should we do to get one?

The procedure carried out above, along with the answer to problem P16, illustrates an algorithm that will always produce a maximal feasible flow in any network with nonnegative integer capacities. (Later we'll discuss networks that contain non-integer capacities.)

P17 Try it on this network, starting with $\phi_1 = (UABV)$.

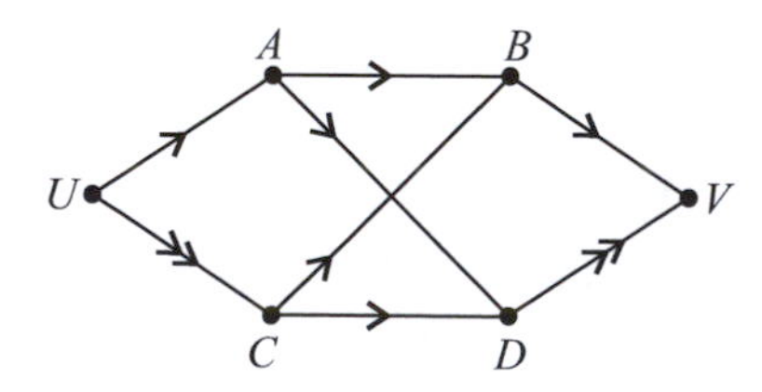

Before stating the maximal flow algorithm in general, we note that often it is possible to save steps by taking the starting flow ϕ_1 to be more than just a single path flow: Any feasible flow in $\mathbf{N}_1$ can be used as ϕ_1. So for example in problem P17, ϕ_1 can be $(UABV) + (UCDV)$. Then the algorithm terminates after only one more step.

P18 Try it.

The maximal flow algorithm

To construct a maximal flow ϕ in a given network $\mathbf{N}$ with nonnegative capacities do this:

First step Let $\mathbf{N} = \mathbf{N}_1$ and start with any feasible flow ϕ_1 in $\mathbf{N}_1$. Reverse all units of capacity used by ϕ_1 to obtain a new network $\mathbf{N}_2$.

General step For any $k > 1$, let ϕ_k be a saturating path flow in any path of positive capacity from the source to the sink in the network $\mathbf{N}_k$. Reverse all units of capacity used by ϕ_k to obtain the next network $\mathbf{N}_{k+1}$.

Stop when no such path exists. Finally, obtain a maximal feasible flow ϕ by (if necessary) reducing $\phi_1 + \phi_2 + \cdots + \phi_n$.

In other words, ϕ is the unique reduced flow that is equivalent to the sum $\phi_1 + \phi_2 + \cdots + \phi_n$. (Refer back to problem P10 if necessary.)

To clarify what is meant by reversing units of capacity used by a flow ϕ, for each arc AB in the network, $c(AB)$ gets reduced by $\phi(AB)$ and the capacity of the reverse arc BA gets increased by the same amount. (Arcs not in the network are considered the same as arcs of capacity zero.)

To prove that the maximal flow algorithm actually does what we claim, we have to show three things:

(1) The process terminates in a finite number of steps;

(2) ϕ is feasible;

(3) ϕ is maximal.

Surprisingly, statement (1) is not necessarily true if all real numbers are allowed as capacities. This problem will not be addressed. For now, we will show that (1) is true under the assumption that all capacities are nonnegative integers. Then it will follow that (1) is also true for networks with rational capacities.

P19 Show that if all arcs of $\mathbf{N}$ have nonnegative integer capacities, then on each step of the algorithm, the value of the flow $\phi_1 + \phi_2 + \cdots + \phi_n$ increases by at least 1, and that this cannot go on forever. (Why not?) Conclude that the algorithm eventually must terminate.

P20 Use the result of problem P19 to show that the same is true for nonnegative rational capacities. (Hint: lowest common denominator.)

To prove that ϕ is feasible, we interpret the algorithm in matrix form. Recall the flow matrix $[\phi]$, for any flow ϕ, and the capacity matrix, which we denote by the symbol $[\mathbf{N}]$, for any network $\mathbf{N}$.

Looking again at this example,

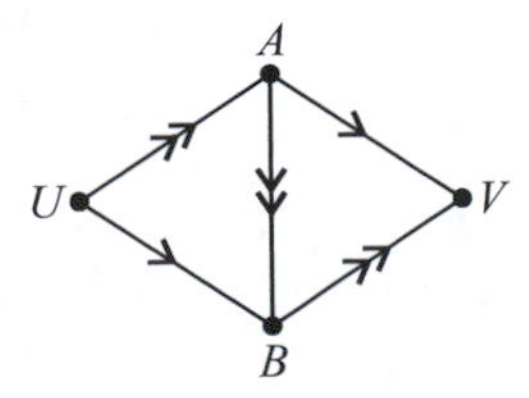

We have the capacity matrix

$$[\mathbf{N}_1] = \begin{bmatrix} 0 & 2 & 1 & 0 \\ 0 & 0 & 2 & 1 \\ 0 & 0 & 0 & 2 \\ 0 & 0 & 0 & 0 \end{bmatrix}$$

and taking $\phi = 2(UABV)$, the flow matrix

$$[\phi_1] = \begin{bmatrix} 0 & 2 & 0 & 0 \\ -2 & 0 & 2 & 0 \\ 0 & -2 & 0 & 2 \\ 0 & 0 & -2 & 0 \end{bmatrix}.$$

P21 Continue this way. Find $[\mathbf{N}_2]$, $[\phi_2]$, and $[\mathbf{N}_3]$. What relationship exists between the matrices?

The observation above suggests that we introduce the following notation: For any network $\mathbf{N}$ with nonnegative capacities and for any feasible flow ϕ in $\mathbf{N}$, the symbol $\mathbf{N} - \phi$ will be used to represent the network obtained from $\mathbf{N}$ by reversing all units of capacity used by ϕ.

P22 $[\mathbf{N} - \phi] =$ ______________. Complete the equation and explain it in terms of the values $c(AB)$, $\phi(AB)$, and $\phi(BA)$ for any arc AB. Assume a fixed ordering of the nodes in $\mathbf{N}$.

Another useful notation is this: For two n-by-n matrices $\mathbf{M}_1$ and $\mathbf{M}_2$, we write $\mathbf{M}_1 \leq \mathbf{M}_2$ to indicate that each element of $\mathbf{M}_1$ is less than or equal to the corresponding element of $\mathbf{M}_2$. Equivalently, all elements of the difference $\mathbf{M}_2 - \mathbf{M}_1$ are nonnegative.

P23 Let ϕ be a reduced flow in a network $\mathbf{N}$. Prove that ϕ is feasible in $\mathbf{N}$ if and only if $[\phi] \leq [\mathbf{N}]$.

In the new notation, the maximal flow algorithm takes this form:

$$\begin{aligned} \mathbf{N}_2 &= \mathbf{N}_1 - \phi_1 \\ \mathbf{N}_3 &= \mathbf{N}_2 - \phi_2 \\ &\vdots \\ \mathbf{N}_{n+1} &= \mathbf{N}_n - \phi_n \end{aligned}$$

As before, ϕ_n is the last path flow produced and the result of the algorithm is the reduced flow ϕ that is equivalent to the sum $\phi_1 + \phi_1 + \cdots + \phi_n$.

P24 Prove that $[\phi] \leq [\mathbf{N}_1]$. Conclude by problem P23 that ϕ is feasible in $\mathbf{N}_1$.

Finally, it remains to show that ϕ is maximal. To do this we introduce the concept of a *cut* in a network. By this we mean a partition of the nodes into two sets $\mathbf{X}$ and $\mathbf{Y}$ having these properties:

X and **Y** are disjoint;
Every node is in either **X** or **Y**;
All source nodes are in **X**;
All sink nodes are in **Y**.

In our case, there is just one source and one sink. The motivation for the term "cut" comes from diagrams like this:

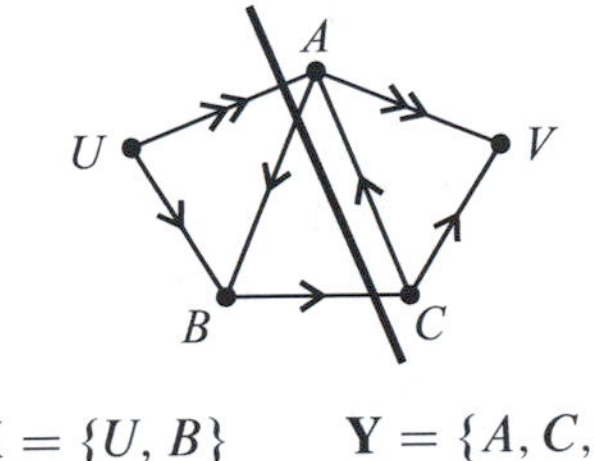

$$\mathbf{X} = \{U, B\} \qquad \mathbf{Y} = \{A, C, V\}$$
$$\text{or} \quad \mathbf{X}/\mathbf{Y} = UB/ACV$$

P25 List all of the cuts that exist in this network, but first predict how many there are. Represent the cut UC/ABV on a diagram like the one above.

Cuts and Capacities

We will look at the relationship between flows and cuts in a network and how the capacities of certain arcs provide information about the maximal flow value.

P26 Return to this example

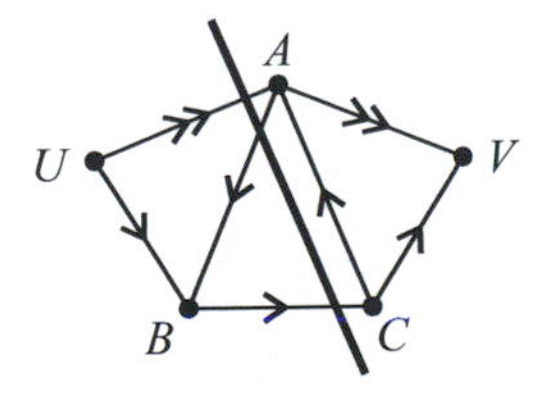

In which arrows represent units of capacity. Taking into account only arcs that cross the "cut line," what can you say about the value of any feasible flow in this network?

P27 Answer the same question using the cut UA/BCV. How does this information compare with the (obvious) maximal flow value in this network?

P28 (a) Make up reasonable definitions for the following: a *forward arc* of a cut; a *backward arc* of a cut; the *capacity of a cut* (notation: $c(\mathbf{X}/\mathbf{Y})$).

(b) Find the capacity of each cut in the network in problem P26. Guess how these numbers relate to the maximal flow value?

P29 Complete the statement: In any network with nonnegative capacities on all arcs, if ϕ is any feasible flow and $\mathbf{X}/\mathbf{Y}$ is any cut, then ____________________________.

P30 State a max-min principle for flows and cuts, similar to those which apply to matching and coverings in a bipartite graph and to antichains and chain decompositions in a cycle-free digraph.

P31 Thinking of the König–Egervary Theorem and Dilworth's Theorem, guess what is true about flows and cuts. (We'll prove this below.)

Now we're ready to nail everything down. We will prove that the maximal flow algorithm actually constructs a maximal flow and that the conjecture in problem **P31** is correct. At the same time, we will see how to obtain a cut of minimal capacity. All of this will become clear by looking in the right place in the flow and capacity matrices.

Starting with the capacity matrix, look back at the network in problem **P26** with the cut UB/ACV.

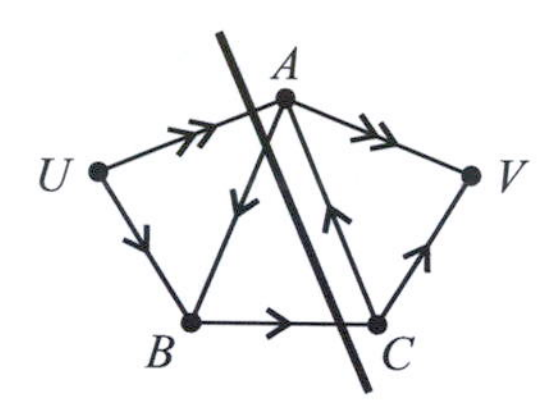

The capacity matrix can be put in this form

	U	B	A	C	V
U	0	1	2	0	0
B	0	0	0	1	0
A	0	1	0	0	2
C	0	0	1	0	1
V	0	0	0	0	0

where we have rearranged the nodes so that all nodes in $\mathbf{X}$ come before all nodes in $\mathbf{Y}$.

P32 Interpret the capacity of the cut UB/ACV as the sum of elements in a submatrix of the capacity matrix. Generalize to any cut in any network.

P33 In the same network, consider the flow $\phi = (UAV) + (UABCV)$. Construct the flow matrix $[\phi]$ relative to the node ordering (U, B, A, C, V).

P34 Referring to the flow ϕ and the matrix $[\phi]$ obtained in the preceding problem:

(a) What is the value of ϕ?

(b) Find the sum of the elements in each row of $[\phi]$. Explain why you got these results.

(c) Explain why the elements in the upper-right 2-by-3 submatrix add up to the flow value $v(\phi)$. (Hint: What must be true about the sum of the elements in the upper-left 2-by-2 submatrix?)

P35 Generalize the result observed in problem **P34**(c) to any flow in any matrix. Look at the upper-right submatrix in the first k rows and last $n - k$ columns, where the total number of nodes is n and the nodes are ordered with the source first and the sink last.

P36 Let ϕ be a feasible flow in a network $\mathbf{N}$, and let $\mathbf{X}/\mathbf{Y}$ be any cut in $\mathbf{N}$. Use matrices to prove that $c(\mathbf{X}/\mathbf{Y}) - v(\phi)$ is equal to the capacity of $\mathbf{X}/\mathbf{Y}$ as a cut in the network $\mathbf{N} - \phi$. (If necessary, look back at what we proved in problem **P22**.)

P37 Directly verify the result of problem P36 for the network, cut, and flow of problems P32–34.

The result in problem P36 is what we need to prove everything we have claimed. Using an obvious notation, we write it in this form:

$$c_{\mathbf{N}}(\mathbf{X}/\mathbf{Y}) - v(\phi) = c_{\mathbf{N}-\phi}(\mathbf{X}/\mathbf{Y}).$$

Now recall that at the end of the maximal flow algorithm, the final network $\mathbf{N}_{n+1}$ contains no path of positive capacity from the source to the sink.

P38 Prove that $\mathbf{N}_{n+1}$ contains a cut of capacity zero. (Hint: Consider the set of all nodes that are reachable from the source by paths of positive capacity.)

P39 Prove that the flow ϕ constructed by the maximal flow algorithm is a maximal flow.

P40 Let ϕ be any feasible flow in a network $\mathbf{N}$. Show that the implication goes both ways: ϕ is maximal if and only if $\mathbf{N} - \phi$ contains a cut of capacity zero.

P41 Complete the statement below and explain why it works.

The minimal cut algorithm

To construct a minimal capacity cut $\mathbf{X}/\mathbf{Y}$ in a given network $\mathbf{N}$, starting with a maximal flow ϕ, do this: ____

____

____

P42 Construct a maximal flow and use it to obtain a minimal cut in this network. As usual, arrows represent units of capacity.

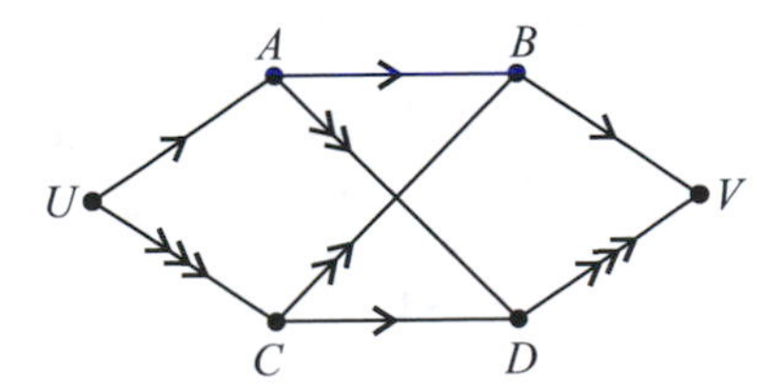

P43 Show that if ϕ is any maximal flow in the network of problem P42, then $\phi(AB)$ must be zero. What can you say about $\phi(CD)$?

As a consequence of what we have done, we have *almost* proved this:

The maxflow/mincut theorem *In any network with nonnegative capacities on all arcs, the maximal flow value is equal to the minimal cut capacity.*

We have proved this under the assumption that a maximal flow exists. If all capacities are rational numbers, then we know that the maximal flow algorithm terminates in a finite number of steps, producing a maximal flow. The problem comes when the network includes irrational capacities. In that case it is still true that a maximal flow exists, and therefore that the theorem above is true in all cases, but we haven't proved it yet.

More Problems

P44 A *circulation* in a network is a flow of value zero. Equivalently, the flow conservation equations apply at all nodes.

(a) Represent this circulation as a sum of cycle flows.

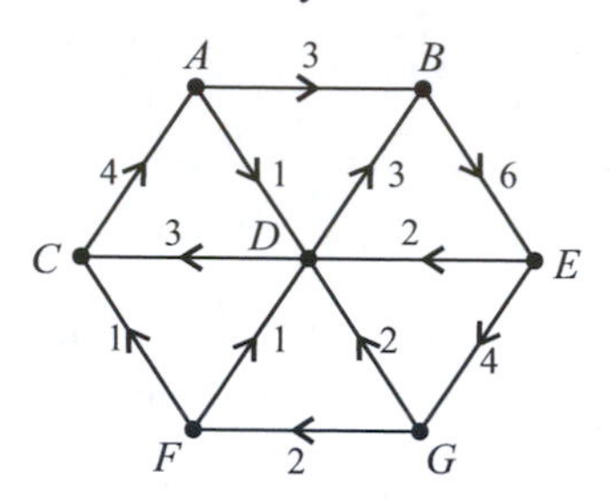

(b) Prove that every circulation can be represented as a sum of cycle flows. (Hint: Induction on the number of arcs containing positive flow.)

P45 Use the maximal flow algorithm to obtain a maximal flow in this network. Arrows represent units of capacity.

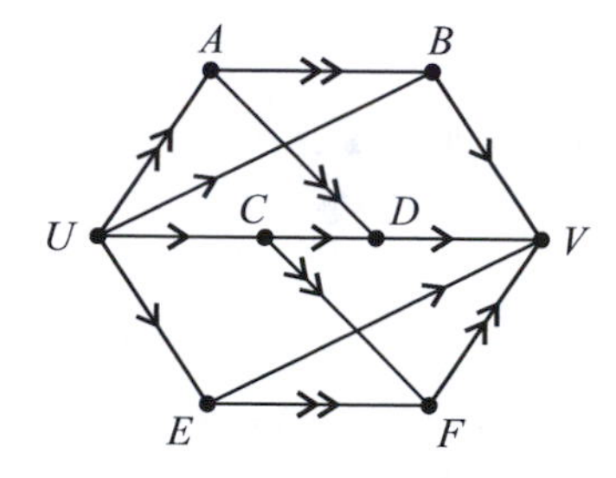

P46 Find a maximal flow in this network:

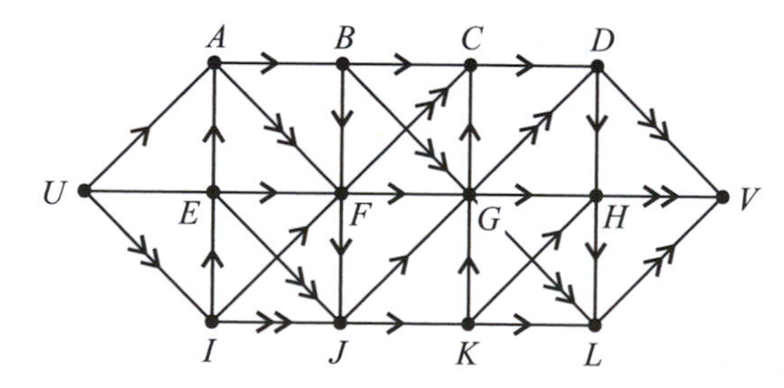

P47 Suppose we want to find a maximal flow in a network with multiple sources and/or sinks. First decide what that means and then indicate how we can solve this problem:

(a) by adding something to the network;

(b) without adding anything to the network.

P48 What's wrong with this argument? In a network with arbitrary nonnegative real capacities, let c_0 be the smallest positive capacity. Then on each step of the maximal flow algorithm, the flow value increases by at least c_0. This cannot go on forever, so the algorithm must terminate in a finite number of steps.

P49 Recall that a flow in which $\phi(AB)$ is an integer for all arcs AB is called an *integral flow*. Suppose that all capacities in a network are nonnegative integers.

(a) Prove that there must be a maximal flow which is integral.

(b) Is every maximal flow in such a network necessarily integral?

(c) What would be true if all capacities are even?

P50 Suppose that ϕ_1 is a feasible flow in a network **N** and ϕ_2 is a feasible flow in the network $\mathbf{N} - \phi_1$. Show that $\phi_1 + \phi_2$ (or the equivalent reduced flow) is a feasible flow in **N**.

P51 In the preceding problem, suppose also that ϕ_2 is maximal in $\mathbf{N} - \phi_1$. Show that $\phi_1 + \phi_2$ (or the equivalent reduced flow) is maximal in **N**.

P52 Use the minimal cut algorithm to obtain a minimal cut in the network in problem P45.

P53 Repeat the preceding problem for the network in problem P46.

P54 In constructing a minimal cut from a maximal flow, why don't we just let **X** consist of all nodes that are reachable from the source by unsaturated paths? Find a counterexample that shows this does not always work.

P55 Suppose that ϕ is a maximal flow in a network **N** and **X**/**Y** is a minimal cut. What can you say about $\phi(AB)$

(a) if AB is a forward arc of the cut?

(b) if AB is a backward arc of the cut?

Prove your assertions by referring to matrices.

P56 Suppose that some of the arcs in a network **N** have infinite capacity, but there is no path of infinite capacity from the source to the sink. Prove that **N** contains at least one cut that has finite capacity. What does this indicate about feasible flow values in **N**?

P57 It is possible to consider *node capacities* as well as arc capacities in a network. By this we mean an upper bound on the total inflow (or outflow) at a given node. For example, in this network, suppose we place a node capacity of 1 at vertex C.

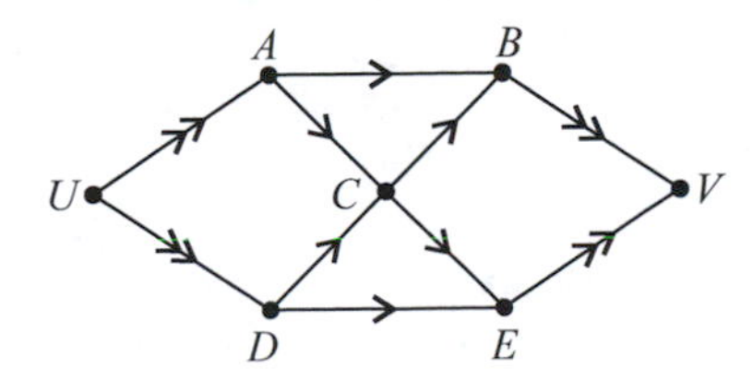

(a) What happens to the maximal flow value?

(b) Explain the result in (a) by converting the network to one with only arc capacities. (Hint: Split C into two nodes C_{in} and C_{out}.)

P58 Suppose that a node capacity of 2 is placed at node G in the network in problem P46. Show that the maximal flow value is reduced to 4.

P59 In a network with nonnegative capacities on all arcs, suppose that there exist arcs of positive capacity going into the source and/or out from the sink. Use the maxflow/mincut theorem to show that there exists a maximal flow that does not use any of these arcs.

P60 In this network, arcs starting at U and arcs ending at V each have capacity 1. Arcs in the center have infinite capacity.

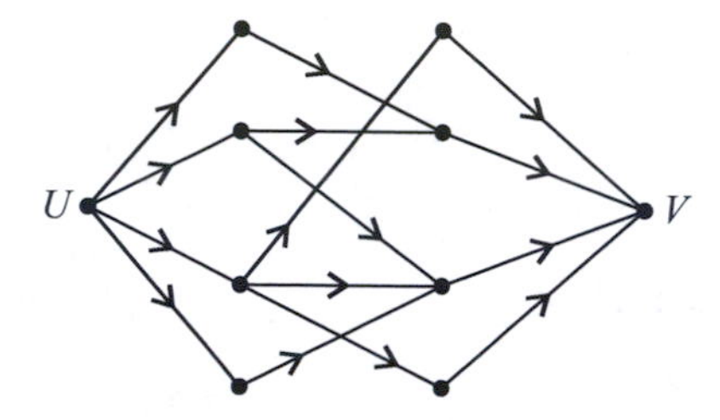

(a) Find a maximal flow and a minimal cut. Locate all of the forward arcs of the minimal cut and explain why none of these occur in the center.

(b) Use the results in (a) to find three independent edges and five independent vertices in this graph.

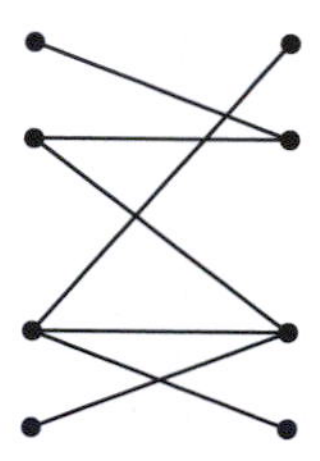

P61 In this graph, the four heavy edges form a maximal matching. Use a minimal cut in the appropriate network to find seven independent vertices.

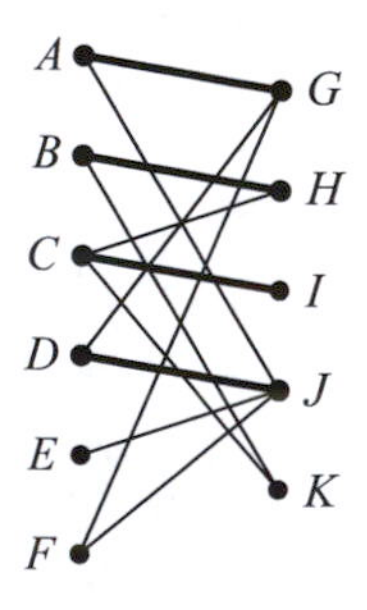

P62 Show how the König–Egervary Theorem can be derived from the maxflow/mincut theorem. Use the notation suggested by this diagram.

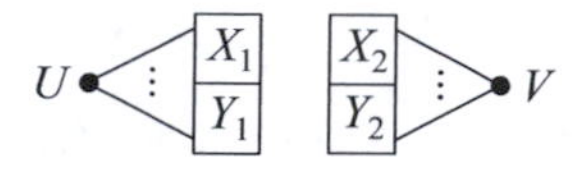

Q

Flow Problems with Lower Bounds

In this section we deal with two closely related problems involving network flows. In the first problem, there are given minimum values for the net inflow at certain nodes. In the second, there are lower capacities as well as upper capacities on arcs. While the first problem is actually a special case of the second, it will be better to solve it separately first.

The Supply and Demand Problem

Let **N** be a network with nonnegative capacities on arcs and any number of sources and sinks, referred to as *supply nodes* and *demand nodes,* respectively. At each supply node P there is an nonnegative quantity $s(P)$, called the *supply* at P, which is an upper limit on the net outflow from P. At each demand node Q, the nonnegative quantity $d(Q)$ is called the *demand* at Q and is a lower limit on the net inflow to Q.

A solution to the supply and demand problem in **N** is a feasible flow that satisfies all demands (net inflow at each demand node Q is at least $d(Q)$) using only the available supplies (net outflow at each supply node P is at most $s(P)$). In general, there is no guarantee that these conditions can be satisfied in a given network. So part of the problem will be to determine whether a solution exists.

Q1 In this network, A and B are supply nodes, and D and E are demand nodes. Arrows indicate units of capacity. Find a solution to the supply and demand problem by trial and error.

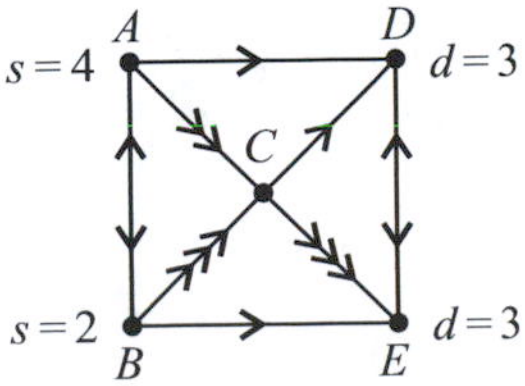

Q2 What does the network **N*** below have to do with the preceding problem? Here U is the source and V is the sink.

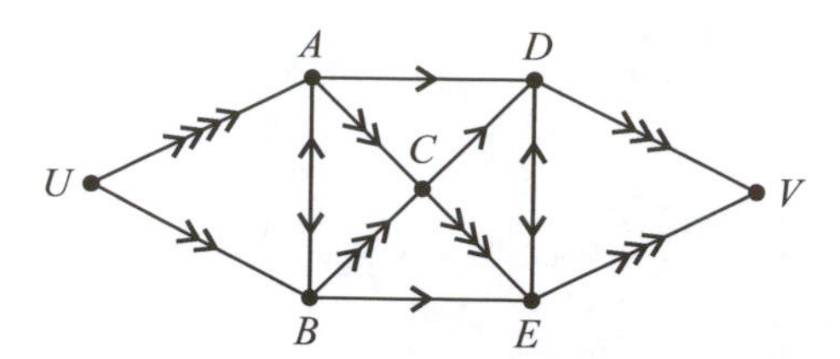

As the problem above suggests, we will solve the supply and demand problem for a given network by applying the maximal flow algorithm to a slightly larger network **N*** in which a source and sink are added. However, there are a few complications that need to be dealt with first. Notice that in the original network **N**, the capacities on the arcs that go to V place an *upper* limit of 3, rather than a *lower* limit, on the net inflow to each demand node in the network of problem Q1. In other words, we are excluding from consideration any flow that overfills one or more of the demands. That's OK in this case, since the total of available supplies is equal to the sum of the demands and consequently there is no chance that a solution might overfill any of the demands. However, it raises the possibility that in some networks, a solution to the supply and demand problem might exist that overfills some demands, while there is no solution that exactly fills them.

Q3 Do you think that could happen?

You're right: That can't happen. But we have to prove it. We begin by making the following definition:

Definition A solution to the supply and demand problem in a given network is *efficient* if the net inflow at each demand node Q is exactly equal to the demand $d(Q)$.

We'll show that if the supply and demand problem has a solution in a given network, then it has an efficient solution. The significance of this fact is that it reduces the problem to finding a maximal flow in another network, a problem we already know how to solve. Specifically, given a supply and demand problem in a network **N**, we construct another network **N*** by adding a source U and a sink V as in problem Q2. Feasible flows in **N*** correspond in an obvious way to flows in **N**.

Supply and Demand Theorem #1

With notation as above, the supply and demand problem in **N** is solvable if and only if **N*** has a feasible flow that saturates all arcs that go to V.

In other words, the condition requires that the maximum feasible flow value in **N*** is equal to the sum of all of the demands in **N**.

The next example illustrates the idea we will use to prove the theorem. Consider this network in which A and B are supply nodes, C and D are demand nodes.

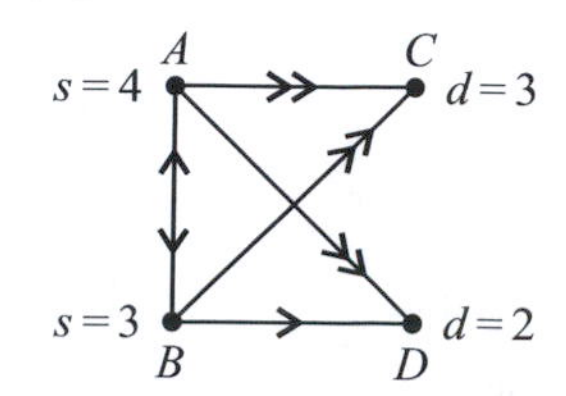

Q4 Solutions to this supply and demand problem, including inefficient solutions, correspond to flows in the network $\mathbf{N}^*$ below. Explain the significance of the capacities on the arcs from U and the arcs to V.

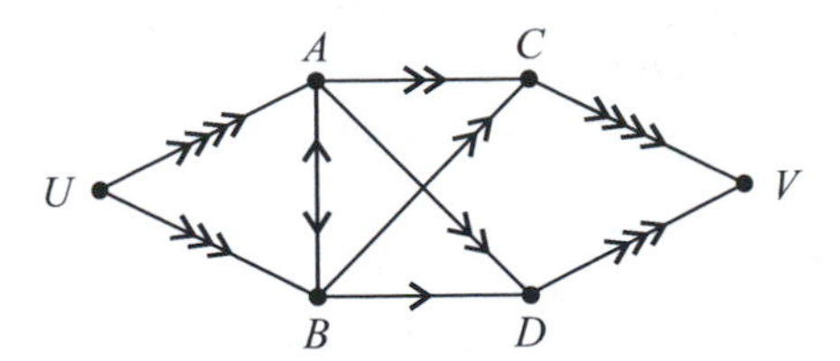

The inefficient solution below at the left corresponds to a flow of value 7, shown at the right, in the network of problem **Q4**. We will show how its existence implies that an efficient solution also exists. The flow is maximal, saturating all arcs that go to V. These arcs are the forward arcs of the cut $UABCD/V$, which is minimal in this network. The minimal cut capacity is 7.

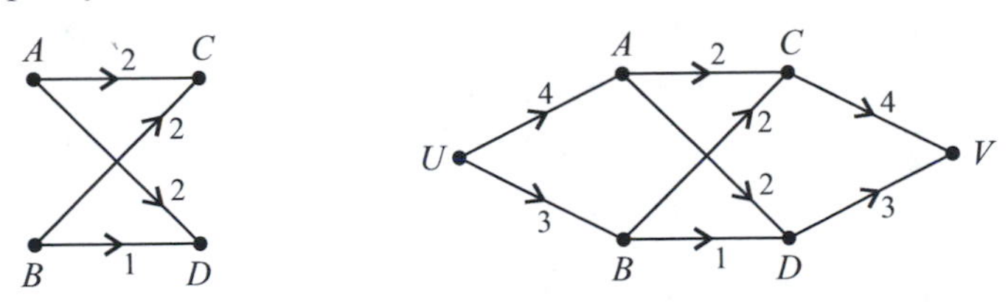

Now we make a slight change in the network, reducing the capacities of the arcs going to V each by one unit. These capacities now match the demands at C and D in the original problem.

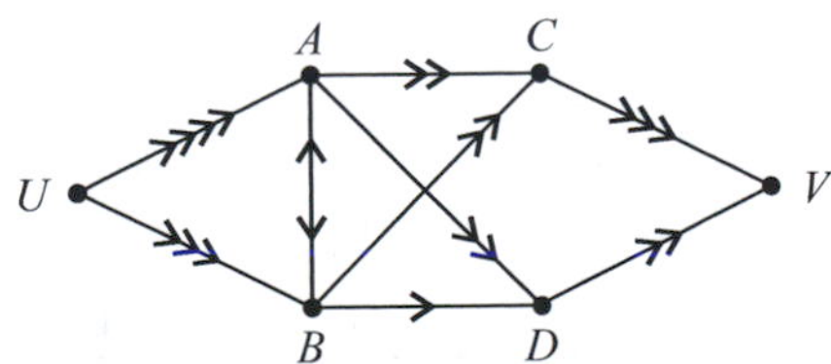

Q5 Show that the cut $UABCD/V$ is still minimal. Do this without actually looking at all of the other cuts individually, but just by considering how the two networks differ. Hint: Is it possible that some other cut could now have capacity less than 5?

Q6 Use the result of problem **Q5** to show that the supply and demand problem in **Q4** has a solution that exactly fills the demands, without actually producing that solution. Hint: Think about maximal flows and minimal cuts.

Problems **Q4–6** illustrate how a solution that overfills some demands can be used to prove the existence of an efficient solution. This is essentially what we need in order to prove Supply and Demand Theorem #1. We now begin the proof in general.

To the original network we add a source U and a sink V, with arcs having the indicated capacities for all supply nodes P and demand nodes Q:

The infinite capacities could be replaced with sufficiently large finite values, such as in problem Q4, but in the general case it's simpler to just start this way.

Now suppose there exists a solution to the supply and demand problem. The net inflow to each demand node Q is $d(Q) + e(Q)$, where $e(Q)$ is a nonnegative quantity representing the extra net inflow beyond the demand. We use this solution to modify the network slightly, reducing capacities on all arcs that go to the sink:

$$Q \xrightarrow[d(Q)+e(Q)]{} V$$

The capacity of QV is now $d(Q) + e(Q)$. Nothing else changes.

Q7 Show that $U \ldots / V$ is a minimal cut in this network. (Hint: maximal flow)

In the next step we change the network one more time, reducing the capacities on arcs going to V still further: Now $c(QV) = d(Q)$, the demand at Q, for each demand node Q. This network is now the one referred to in the theorem as $\mathbf{N}^*$.

Q8 Show that $U \ldots / V$ is still a minimal cut in this latest network. Use this fact to conclude that any maximal flow must saturate all arcs that go to V. What does this have to do with the existence of an efficient solution to the supply and demand problem?

That completes the proof of Supply and Demand Theorem #1.

Q9 In the network below, A, B, and C are supply nodes. F and G are demand nodes. Show that this supply and demand problem has no solution by finding a maximal flow in an appropriately expanded network $\mathbf{N}^*$.

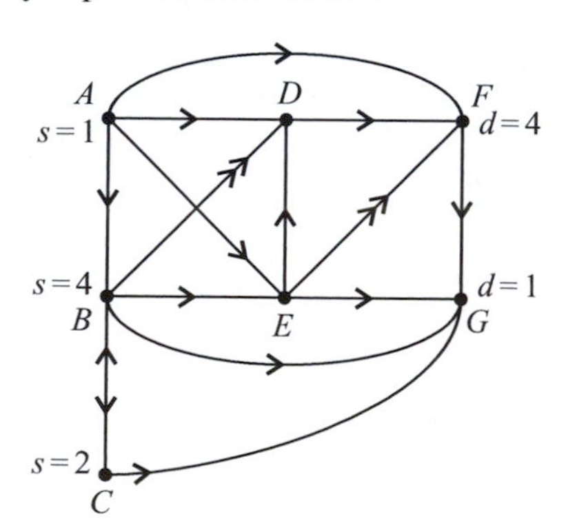

Q10 Find a minimal cut $\mathbf{X}/\mathbf{Y}$ in th expanded network from problem Q9.

Q11 Let $\mathbf{Z}$ be the set of nodes obtained by removing the sink V from the set $\mathbf{Y}$. Find the *net demand* in $\mathbf{Z}$, defined as total demand minus total supply at nodes of $\mathbf{Z}$. Compare this value with the total capacity of all arcs that enter $\mathbf{Z}$ in the original network. Use this to give a new explanation of why no solution can exist.

Problem Q11 suggests the following necessary condition for a given supply and demand problem to have a solution:

For every node set $\mathbf{Z}$ in the network, the net demand in $\mathbf{Z}$ cannot exceed the total capacity of arcs that enter $\mathbf{Z}$.

All of this takes place in the original network, before U and V are added. However, this condition is closely related to cut capacities in the expanded network. In fact, we will use this relationship to prove that the condition above is not only necessary, but also sufficient, for the existence of a solution.

Supply and Demand Theorem #2 (Gale's Feasibility Theorem)

The supply and demand problem is solvable in a given network **N** if and only if, for every set **Z** of nodes in **N**, the net demand in **Z** is less than or equal to the total capacity of all arcs that enter **Z**.

This includes the case in which **Z** is the empty set, but it doesn't really matter since the condition is always satisfied when **Z** is empty.

Q12 What does the condition say when **Z** consists of all nodes of **N**?

To prove the theorem, we let **N*** denote the expanded network obtained from **N** by adding a source U, a sink V, and arcs with the indicated capacities for all supply nodes P and demand nodes Q:

$$U \xrightarrow{s(P)} P \qquad Q \xrightarrow{d(Q)} V$$

N* is the network from Supply and Demand Theorem #1. We know that a solution exists if and only if a maximal flow in **N*** saturates all arcs QV.

For any node set **Z** in **N**, let **Z′** denote the complementary node set in **N**. Let $c(\mathbf{Z}'/\mathbf{Z})$ denote the total capacity of all arcs going from **Z′** to **Z**, and let $s(\mathbf{Z})$, $s(\mathbf{Z}')$, $d(\mathbf{Z})$, $d(\mathbf{Z}')$ represent the total supply or demand in each set. A typical cut $\mathbf{X}/\mathbf{Y}$ in **N*** is obtained by adding U to **Z′** and adding V to **Z**.

Q13 Complete the equation:

$$c(\mathbf{X}/\mathbf{Y}) = c(\mathbf{Z}'/\mathbf{Z}) + \underline{\qquad\qquad}.$$

Q14 Show that $c(\mathbf{X}/\mathbf{Y}) \geq d(\mathbf{Z}) + d(\mathbf{Z}')$ if and only if $c(\mathbf{Z}'/\mathbf{Z}) \geq d(\mathbf{Z}) - s(\mathbf{Z})$. How does this prove the theorem?

The lower capacity problem

We solved the maximal flow problem in chapter P for networks in which arcs have upper capacities: that is, the capacity $c(AB)$ is an upper limit on $\phi(AB)$ for each arc AB. Now we consider what happens when we also place lower limits on $\phi(AB)$. We will write $\mathrm{lc}(AB)$ and $\mathrm{uc}(AB)$, respectively, to denote the lower capacity and upper capacity of a given arc. Previously, $\mathrm{lc}(AB)$ was always zero. Now we will allow positive lower capacities, assuming that for any arc AB, $0 \leq \mathrm{lc}(AB) \leq \mathrm{uc}(AB)$.

The *lower capacity problem* in a given network is actually two problems: First determine whether a feasible flow exists; this requires that

$$\mathrm{lc}(AB) \leq \phi(AB) \leq \mathrm{uc}(AB)$$

for each arc. And then, if it does, find a maximal feasible flow. We'll consider the existence problem first.

Q15 In this network, arrows → represent units of lower capacity.

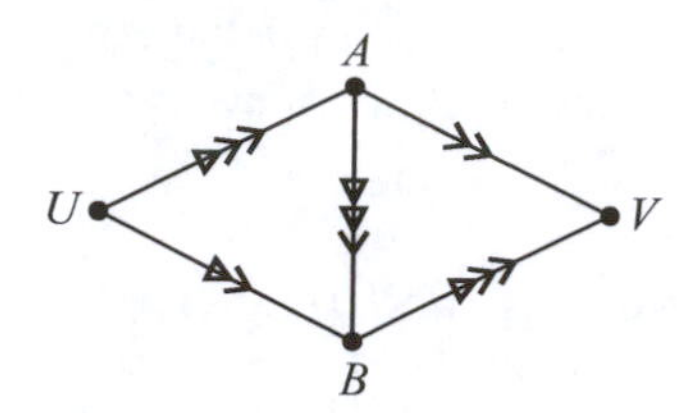

For example, arc UA has lc $= 1$ and uc $= 3$. Find a feasible flow by trial and error. (As usual, consider U and V to be the source and sink.)

In developing a systematic approach to the problem above, we begin by adding arc VU to the network with infinite upper capacity and lower capacity zero.

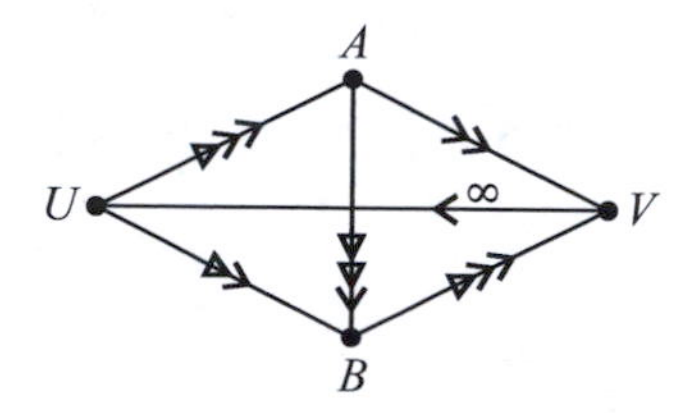

The purpose of the extra arc is to convert U and V—temporarily—to ordinary nodes, at which the flow conservation equations are satisfied. Any feasible flow in the original network will correspond in an obvious way to a *circulation*, which is a flow of value zero, in the expanded network. Any feasible circulation ϕ produces a new inflow or outflow of zero at each node.

Now define a function α on the arcs of the new network by the formula $\alpha = \phi - \text{lc}$:

$$\begin{aligned}
\alpha(UA) &= \phi(UA) - 1 \\
\alpha(UB) &= \phi(UB) - 1 \\
\alpha(AB) &= \phi(AB) - 2 \\
\alpha(AV) &= \phi(AV) \\
\alpha(BV) &= \phi(BV) - 1 \\
\alpha(VU) &= \phi(VU)
\end{aligned}$$

α represents the extra quantity of flow beyond the lower capacity that a solution ϕ sends through each arc. Our strategy is to find ϕ by first finding α. The following problems show how this is done.

Q16 Find the net inflow or outflow produced by α at each node. For example, at U there is a net outflow of

$$\begin{aligned}
&\alpha(UA) \quad + \alpha(UB) \quad - \alpha(VU) \\
&= \phi(UA) - 1 \quad + \phi(UB) - 1 \quad - \phi(VU) = -2,
\end{aligned}$$

or a net inflow of 2. (Why do the other terms cancel off?)

Q17 What does this supply and demand problem have to do with the preceding problem?

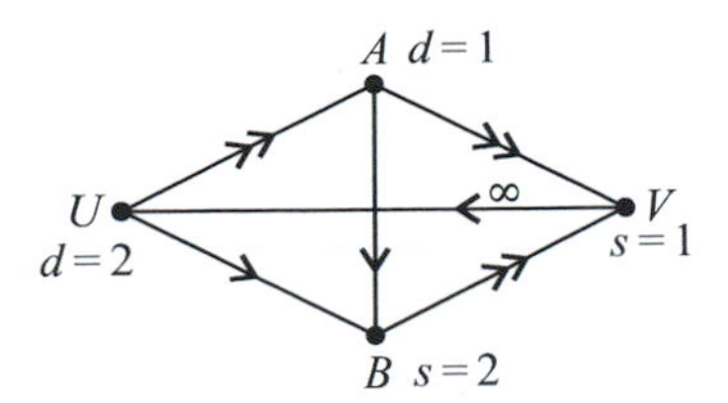

Q18 Find a solution ϕ to the supply and demand problem above.

Q19 Use your answer to problem **Q18** to obtain a feasible flow in the network of problem **Q15**. Compare with your answer to **Q15**. Where do the capacities in problem **Q17** come from? Where do the supply and demand numbers come from?

In general, the problem of finding a feasible flow in a network containing lower capacities can always be converted to a supply and demand problem. Arcs UV and VU are added to the network with infinite (or sufficiently large) capacities in order to convert feasible flows into circulations. This example

shows why UV might be needed. If UV and/or VU are already in the network, then their upper capacities are increased to infinity.

Q20 Show that no feasible flow exists in this network by looking at the appropriate supply and demand problem. U and V are the source and sink.

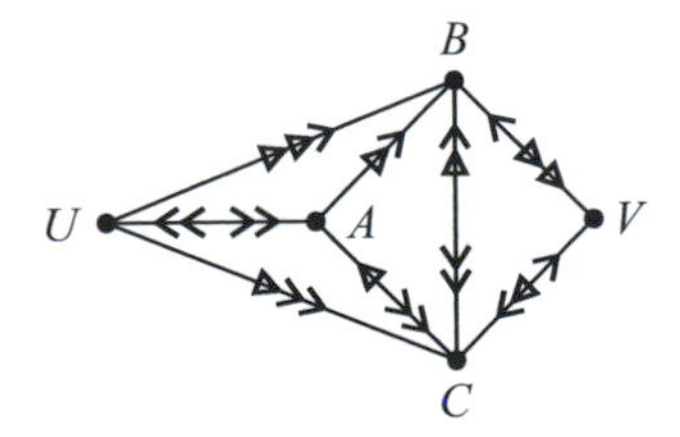

Maximizing the flow

Next we consider the problem of finding a maximal flow in a network with both upper and lower capacities. We assume that a feasible flow ϕ_0 exists and is already known. We'll see how a maximal flow can be obtained starting with ϕ_0 and making a few small changes in the maximal flow algorithm.

As a first example, we start with the flow $\phi_0 = 2(UABV)$ in this network.

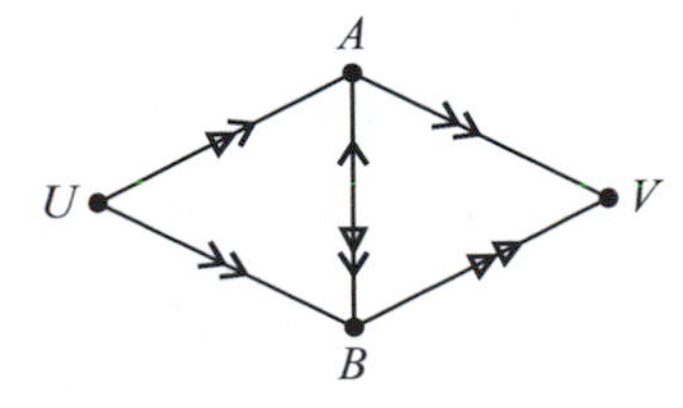

The usual maximal flow algorithm would produce the next network $\mathbf{N} - \phi_0$ by reversing all units of capacity used by ϕ_0. Now, instead, we reverse only optional units of capacity used by ϕ_0 and *remove* all required units.

Q21 Construct $\mathbf{N}_1 = \mathbf{N} - \phi_0$ and guess what to do next.

Q22 The algorithm terminates after ϕ_1. But $\phi_0 + \phi_1$ is not feasible in $\mathbf{N}$ and neither is the reduced flow obtained from $\phi_0 + \phi_1$. Think of something else.

This example suggests the following general procedure: Starting with a feasible flow ϕ_0, form $\mathbf{N}_1 = \mathbf{N} - \phi_0$ as above and continue with the usual maximal flow algorithm. However at the end, instead of just reducing the flow $\phi_0 + \phi_1 + \phi_2 + \cdots + \phi_n$ to obtain a feasible flow, we try to adjust it by adding or subtracting 2-cycle flows. In the example above, the 2-cycle flow (ABA) had to be subtracted. The obvious question here is whether such an adjustment is always possible. Fortunately, it is, as we will prove.

The following definition will be useful: In a given network $\mathbf{N}$ with lower capacities, a flow ϕ is called $\mathbf{N}$-*reduced* if for each arc AB, $\phi(AB) \geq \text{lc}(AB)$, $\phi(BA) \geq \text{lc}(BA)$, and equality holds in at least one of these. (In case arc BA doesn't exist in $\mathbf{N}$, we can still consider it to exist with capacities, both upper and lower, equal to zero.)

Q23 Recall that two flows are *equivalent* if and only if they differ by 2-cycle flows. Prove this lemma:

Lemma. *In any network* $\mathbf{N}$ *with lower capacities, every flow is equivalent to a unique* $\mathbf{N}$*-reduced flow.* (Note that an $\mathbf{N}$-reduced flow is not necessarily feasible in $\mathbf{N}$ since the upper capacities may not be satisfied.)

Q24 Find the $\mathbf{N}$-reduced flow that is equivalent to this one in the network preceding problem Q21.

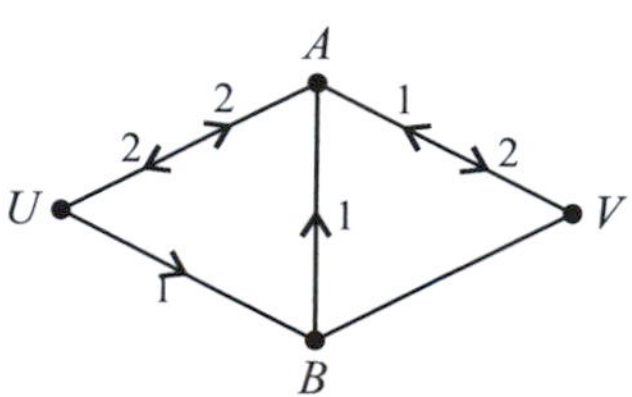

Maximal flow algorithm in networks with lower capacities

Let $\mathbf{N}$ be a network with capacities on all arcs satisfying $0 \leq \text{lc} \leq \text{uc}$. To construct a maximal feasible flow in $\mathbf{N}$, do this:

The extended maximal flow algorithm Start with any feasible flow ϕ_0 in $\mathbf{N}$ and form the network $\mathbf{N}_1 = \mathbf{N} - \phi_0$ by removing all required units of capacity and reversing all optional units of capacity used by ϕ_0. Apply the usual maximal flow algorithm to the network $\mathbf{N}_1$, obtaining a sequence of path flows $\phi_1, \phi_2, \ldots, \phi_n$. Finally, let ϕ be the $\mathbf{N}$-reduced flow that is equivalent to the sum $\phi_0 + \phi_1 + \phi_2 + \cdots + \phi_n$. Then ϕ is a maximal feasible flow in $\mathbf{N}$.

Q25 Use the algorithm to produce a maximal feasible flow in this network.

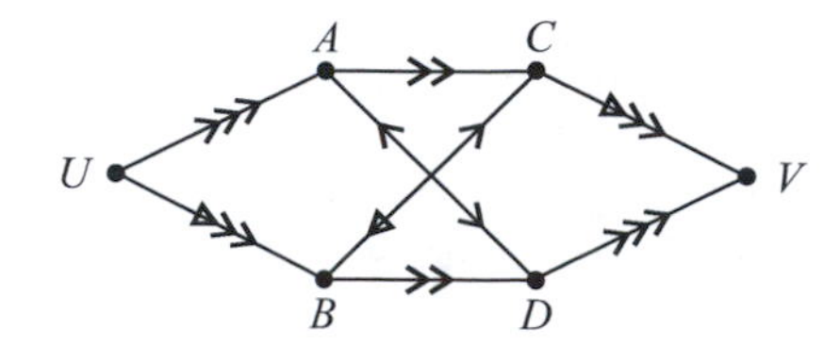

Q26 In the network above, look for a cut that has capacity equal to the value of the maximal flow that you found. Consider redefining the capacity of a cut in a network with lower capacities: What should the definition be?

Like the original maximal flow algorithm, this one can be carried out in matrix form. Matrices will also be used to prove that the algorithm works. However, we have to decide how to define [**N**] when **N** contains both upper and lower capacities.

Look again at this network.

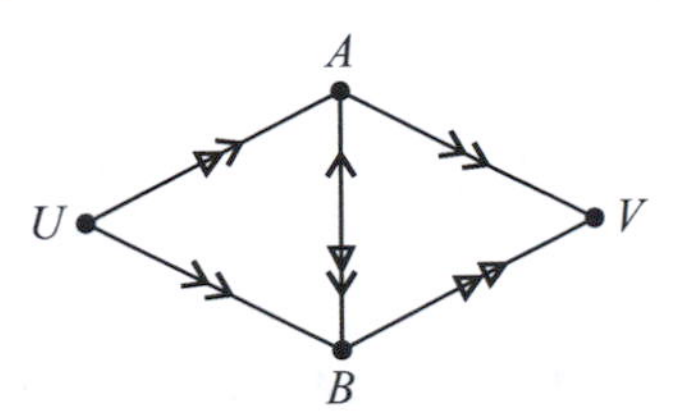

Starting with the feasible flow $\phi_0 = 2(UABV)$, we obtained the network $\mathbf{N}_1 = \mathbf{N} - \phi_0$ shown below. $\mathbf{N}_1$ contains only upper capacities.

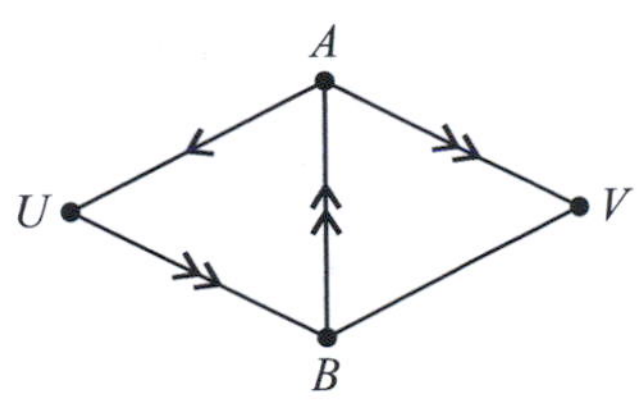

Q27 We would want to define [**N**] so that $[\mathbf{N} - \phi_0] = [\mathbf{N}] - [\phi_0]$. We know $[\mathbf{N} - \phi_0]$ and $[\phi_0]$, so we also know [**N**]. What is it?

Q28 Compare the network **N** with the matrix [**N**]. How could we have gone directly from **N** to [**N**]? Generalize to any network with both upper and lower capacities. What goes into the matrix in the row corresponding to node A and the column corresponding to node B?

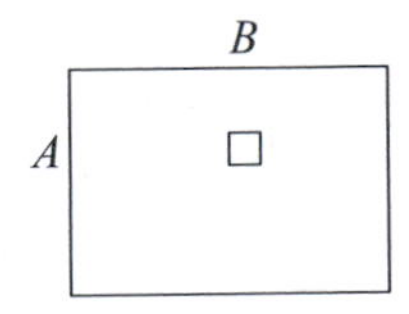

Q29 For any flow ϕ in any network, answer the same question for the flow matrix $[\phi]$. What goes into the AB position?

Q30 Use your answers to problems Q28 and Q29 to show that if ϕ is a feasible flow in **N**, then $[\phi] \leq [\mathbf{N}]$.

Q31 Prove that if ϕ is an **N**-reduced flow and $[\phi] \leq [\mathbf{N}]$, then ϕ is feasible in **N**.

Now we can prove that the extended maximal flow algorithm always produces a maximal feasible flow.

Q32 Use matrices to prove that ϕ is feasible, where ϕ is the flow produced by the algorithm.

Finally, the proof that ϕ is maximal is essentially the same as for the original maximal flow algorithm: We produce a cut in **N** having capacity equal to the value of ϕ. There are just a few details to clean up, since we now have a new definition of cut capacity.

Q33 Fill in the capacity matrix [**N**] for this network.

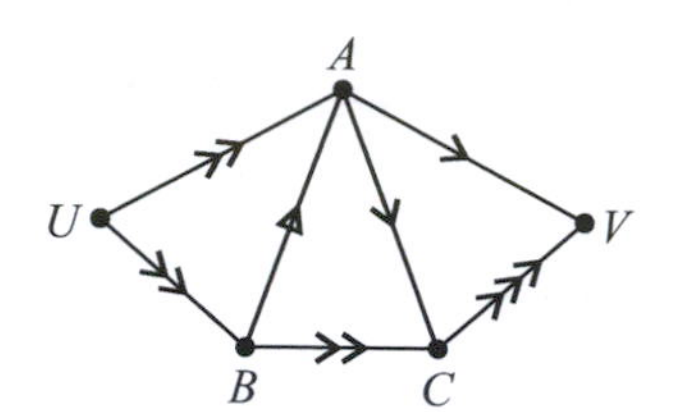

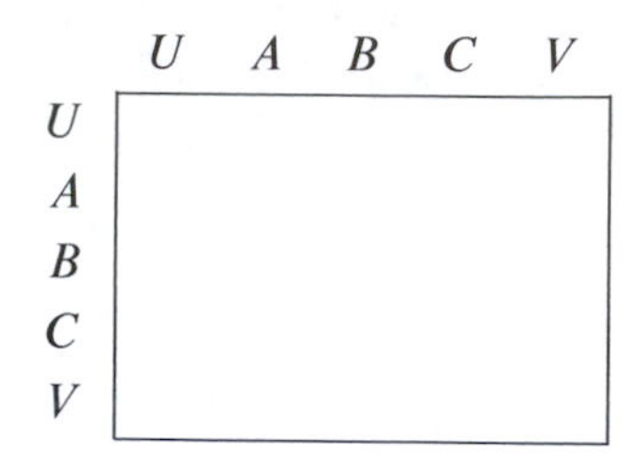

Q34 In the network above, find $C(UA/BCV)$ and compare the result with the sum of elements in a submatrix of [**N**]. Explain in terms of matrices why any feasible flow in **N** must have value $\leq c(UA/BCV)$.

Q35 Find a maximal flow in the network of problem Q33.

Q36 Returning now to the general case of a network with upper and lower capacities, let ϕ be the flow produced by the extended maximal flow algorithm.

(a) Show that $\mathbf{N} - \phi$ contains a cut of capacity zero.

(b) Show that the same cut, in **N**, has capacity equal to the value of ϕ.

(c) What does this prove about ϕ?

More Problems

Q37 (a) Solve this supply and demand problem by trial and error.

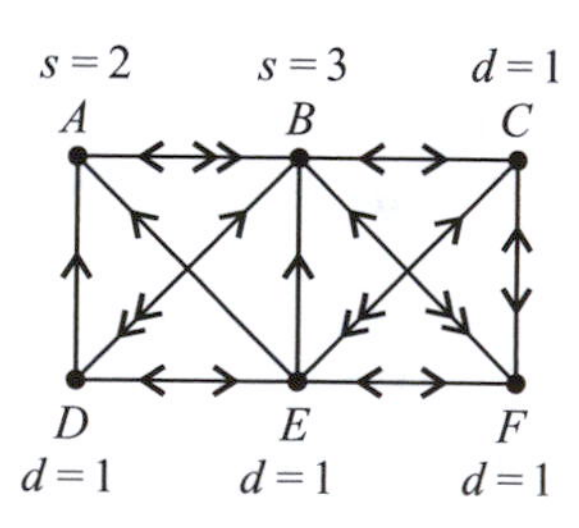

(b) Show how the solution corresponds to a maximal flow in an expanded network $\mathbf{N}^*$.

Q38 Show that there does not exist a bipartite graph having six vertices on each side with degrees 5, 4, 4, 1, 1, 1 by interpreting the problem as a supply and demand problem.

Q39 Show that there does not exist a directed graph having six vertices with indegrees 5, 4, 4, 1, 1, 1 and the same sequence of outdegrees (in any order, so that the vertex with indegree 5 doesn't have to be the same vertex as the one with outdegree 5).

Q40 Use Gale's Feasibility Theorem to prove that this supply and demand problem has no solution.

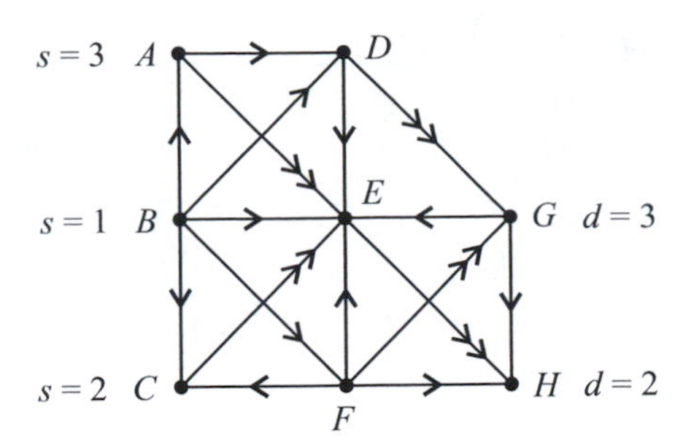

Q41 Suppose that a supply and demand problem has a solution that doesn't use up all of the supply at a particular supply node P. Prove that there exists a solution in which no flow occurs on arcs that go to P. (Suggestion: Show that these arcs can't be forward arcs of a minimal cut in the expanded network $\mathbf{N}^*$ containing U and V.)

Q42 For this network,

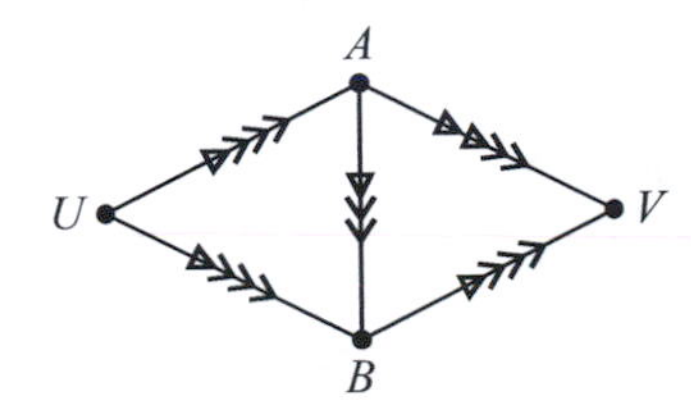

(a) Find a feasible flow ϕ_0 by trial and error.

(b) Show how ϕ_0 can be obtained by solving a supply and demand problem.

(c) Use the extended maximal flow algorithm to construct a maximal flow.

(d) Find a minimal cut in this network.

Q43 Suppose that in the network of problem P45, a lower capacity of 1 is placed on arc CD. The upper capacity is raised to 2. Show that the maximum flow value is reduced to 3.

Q44 Suppose that in the network of problem P46, the capacities on arcs BF and GC are changed to $\text{lc} = 1$, $\text{uc} = 2$. What happens to the maximal flow value?

Q45 For this network $\mathbf{N}$,

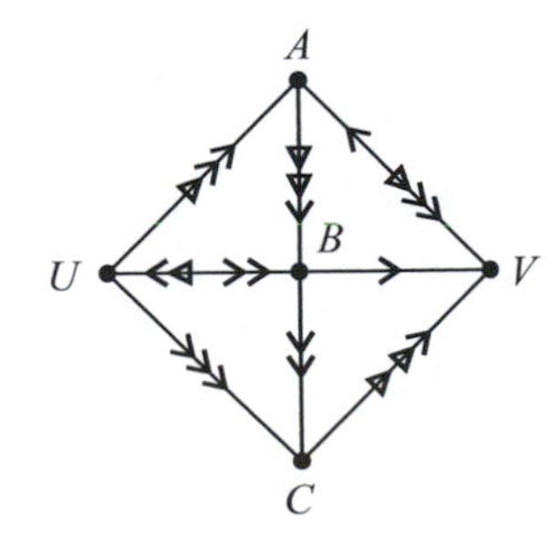

(a) Find a feasible flow ϕ_0;

(b) Construct the network $\mathbf{N} - \phi_0$;

(c) Find a maximal flow;

(d) Find a minimal cut.

Answers to Selected Problems

Chapter A

A2 (a) yes

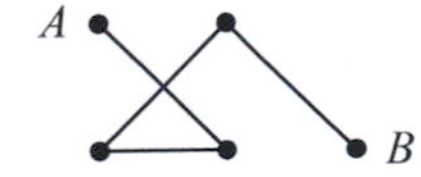

(b)

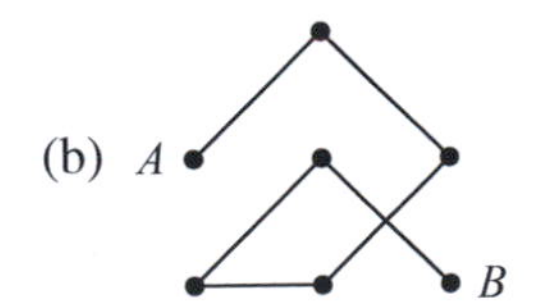

(c) Shortest paths are always simple, but simple paths are not always shortest.

A3 Any shortest path from A and B is simple.

A5 No, because the first vertex is repeated at the end.

A6 A cycle graph consists of a closed path in which the only repeated vertex is the first and last.

A9 $d_1(A) + d_2(A) = n - 1$.

A13 The number of odd vertices must be even.

A15 (b) The degree sum is at most $3 + 4 + 5 + 5 + 5 + 5 + 5 = 32$, so the number of edges is at most 16.

A17 (c) The vertex of degree 6 must be adjacent to all others, but one vertex has degree 0.

A21 Disconnected

A22 (c) path from A to B

Chapter B

B1 no, yes

B3 F T T F T

B4 The vertex of degree 3 is adjacent to two vertices of degree 1 in one graph but to only one in the other.

B8 The first graph contains no cycles of length 3. The second one does.

B10 Graphs 1 and 3 have complements consisting of two components, each of which is C3. The others have complements C6. Therefore 1 and 3 are isomorphic, while 2,4,5, and 6 are isomorphic.

Chapter C

C1 All except the third.

C5 A bipartite graph with m vertices in X and n vertices in Y has at most mn edges. But if $m + n = 17$, then mn is at most 72.

C7 The degree sum in X is equal to the degree sum in Y. Each sum is equal to the number of edges.

C12 They are in different rows and different columns.

Chapter D

D4 $v = e + k$

D5 Clearly the remaining graph **G** contains no cycles. To show that G is connected, let A and B be two vertices in **G**. The original tree contains a simple path from A to B. Being simple, this path cannot pass through a vertex of degree 1 Therefore it exists in **G**.

D6 At an endpoint of a longest simple path **P**, there is one edge in **P**. If there were another edge at that vertex, it would create either a cycle or a longer simple path. Neither can occur, so each endpoint of **P** has degree 1.

D15 (a) There is a path to that point from R, so the indegree is at least 1.

(b) If this were not true, then the sum of all the indegrees would be at least n, the number of vertices in the tree. But this is impossible since there are only $n - 1$ edges.

D19 Each spanning tree in $\mathbf{K}_5$ contains 4 of the 10 edges in $\mathbf{K}_5$. Therefore edge EC has a 40% chance of being in any particular spanning tree. Equivalently, 40% of the spanning trees in $\mathbf{K}_5$ contain this edge. The remaining 60% exist in the given graph. This number is 75.

D21 $AEDBBCCE$

D23 A vertex of degree d occurs d times in the long codeword and $d-1$ times in the short codeword.

Chapter E

E1 $e-1, e-v+1$

E2 There are four of them, each with total weight 4. One is

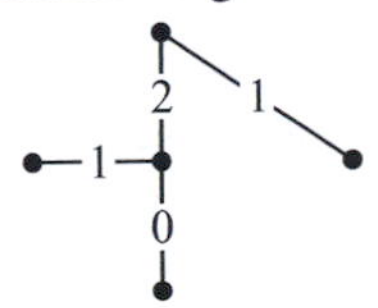

E4 BE and CD

E5 One possibility: Remove BC, AC, and DE.

E8 $AE, AED, AEC, AECB$.

E10 The result is

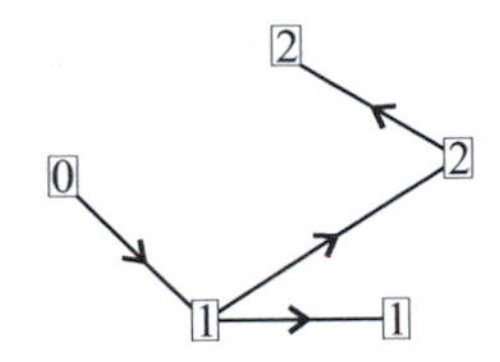

E11

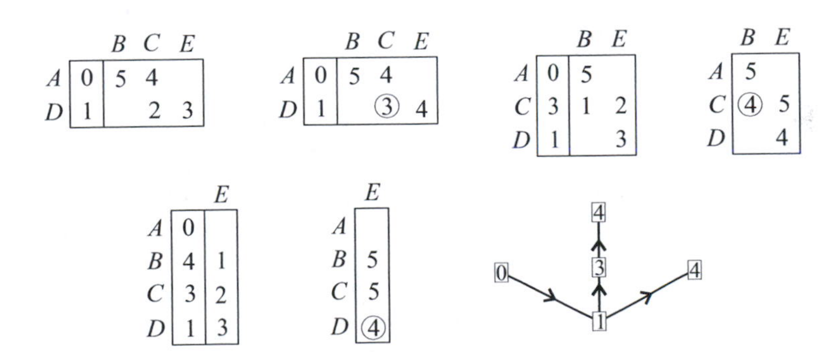

		B	C	E
A	0	5	4	
D	1		2	3

		B	C	E
A	0	5	4	
D	1		③	4

		B	E
A	0	5	
C	3	1	2
D	1		3

	B	E
A	5	
C	④	5
D		4

		E
A	0	
B	4	1
C	3	2
D	1	3

	E
A	
B	5
C	5
D	④

E13 Add ED and EF in either order.

E14 (a) (b)

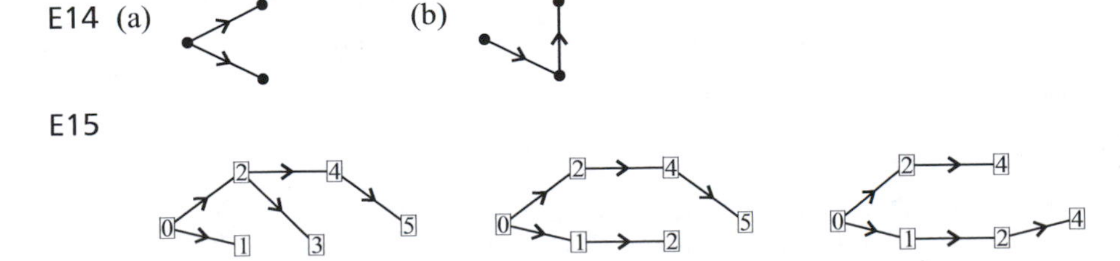

E15

E18 The only minimal path from A to B is $ACDB$. But the only tree that contains this path is the path itself, which doesn't contain a minimal path from A to C.

Chapter F

F2 An odd vertex can occur only at an endpoint of an open Euler path.

F3 (a) 0 (b) 2

F8 A graph with no cycles is a forest. Any component that contains an edge is a tree with more than one vertex. Such a tree contains vertices of degree 1.

F12 (1a) the indegree is equal to the outdegree.
(2a) the indegree is 1 more than the outdegree;
the outdegree is 1 more than the indegree;
the indegree is equal to the outdegree.

F13

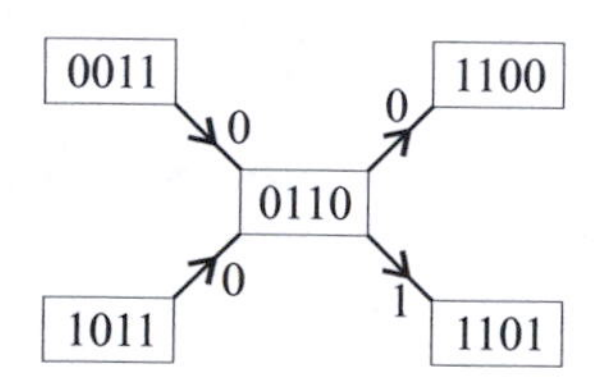

F17 The path includes this, for some x and y:

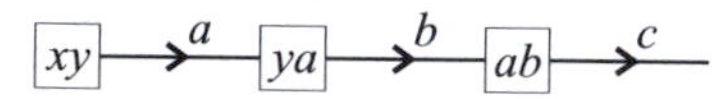

Chapter G

G1 Figure (a) contains a Hamilton path. Figures (c) and (e) contain both.

G5 The first graph contains a Hamilton path.

G9 (a) at most $k+1$ (b) at most k

G12 (a) at most $k+1$ (b) at most k

G14 Removing the white vertices leaves 3 components in (a), 4 in (b), 5 in (c), and 3 in (d).

(a) 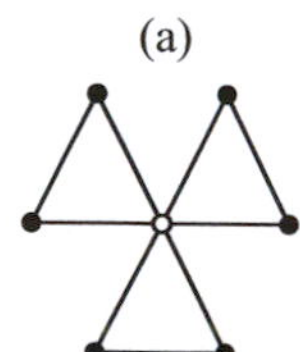

(b) 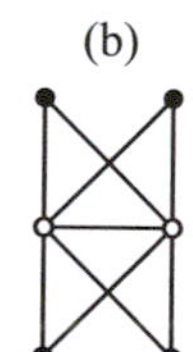

(c) 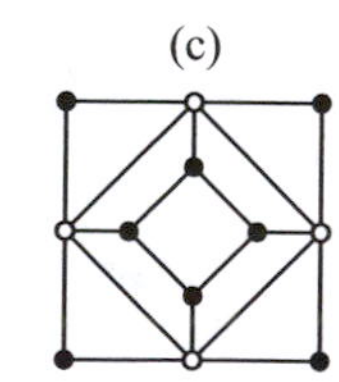

(d) 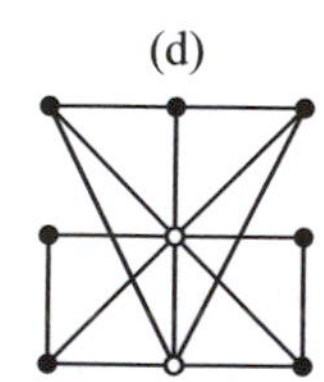

Conclusion: (a) and (b) contain no Hamilton path, (c) and (d) contain no Hamilton cycle.

G16 (b), (c), (e), (f)

G22 $i = 1$ leads to the Hamilton cycle $AV_1BV_6V_5V_4V_3V_2A$
$i = 4$ leads to the Hamilton cycle $AV_1V_2V_3V_4BV_6V_5A$

G24 A is adjacent to V_{i+1}; B is adjacent to V_i. We have to prove these sets have at least one element in common.

G25 Each vertex adjacent to A, except for V_1, determines an element of S. Each vertex adjacent to B, except for V_{n-2}, determines an element of T.

G29 If not, then the same cycle would exist in $\mathbf{G}_k$.

G30 The sum is at least n.

Chapter H

H2 Yes

H4 Add edges to this Hamilton cycle: AC and BF inside the hexagon, DE outside.

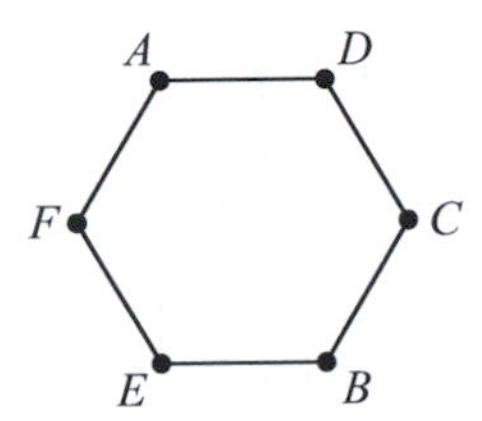

H8 3, 4, 7; 4, 5, 5.

H9 The sum is twice the number of edges. An edge that separates two regions gets counted once for each region. A nonseparating edge gets counted twice for one region.

H12 Euler's Formula shows that there would have to be 7 regions, and each regional degree is greater than or equal to 3.

H16 v remains unchanged while e and r each decrease by 1. Therefore $v - e + r$ remains unchanged.

H18 Euler's Formula shows that the number of edges is 30. If each face has x edges, then the Regional Degree Theorem says that $12x = 60$. Therefore $x = 5$.

H21 $\mathbf{H}_1$ is a subdivision of $\mathbf{K}_5$. $\mathbf{H}_2$ is a subdivision of $\mathbf{K}_{3,3}$. Either of these show that $\mathbf{G}$ is nonplanar.

H22 (a) No subgraph contains vertices of degree 4.
(b) The remaining graph would be planar.
(c) This is a subdivision of $\mathbf{K}_{3,3}$.

Chapter I

I1 α and α' only: α 4 3 2 4 3 9 α' 3 3 3 2 3 4

I2 (b) $n/2$

I5 (a) $2e$
(b) $\alpha d \le e$
(c) $2\alpha d \le 2e = nd$, so $2\alpha \le n$

I7 (b) From (a), $\alpha \ge 10$ and $\alpha' \ge 7$. But $\alpha + \alpha' \le n = 17$, so equality must hold.

I9 The added vertices are all distinct because they come from different members of $\mathbf{S}'$.

I11 Independence Theorem #1 and the corollary to Independence Theorem #2 show that α and α' are each at most $n/2$. Then Independence Theorem #3 implies that both must be $n/2$.

I16 β and β' only: β 3 4 2 4 4 β' 4 3 4 4 9

I17 $\alpha + \beta = n,\ \alpha' + \beta' = n.$

I20 There are $n - 2\alpha'$ vertices that are uncovered by $\mathbf{S}'$. One edge is added to $\mathbf{S}'$ for each of them, and the maximality of $\mathbf{S}'$ implies that these added edges are all distinct. The result is a set of $\alpha' + (n - 2\alpha') = n - \alpha'$ edges.

Chapter J

J2

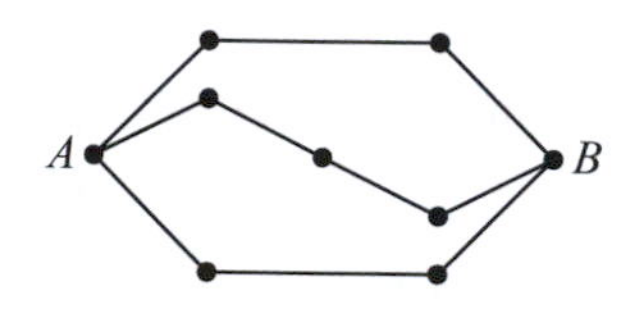

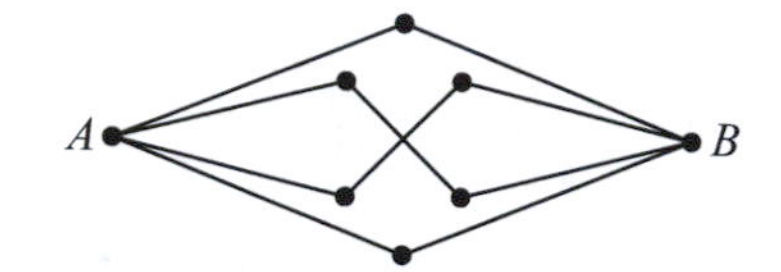

J3

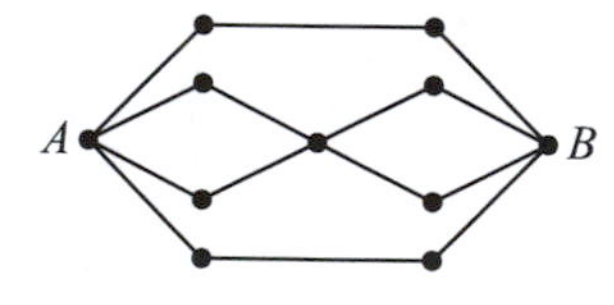

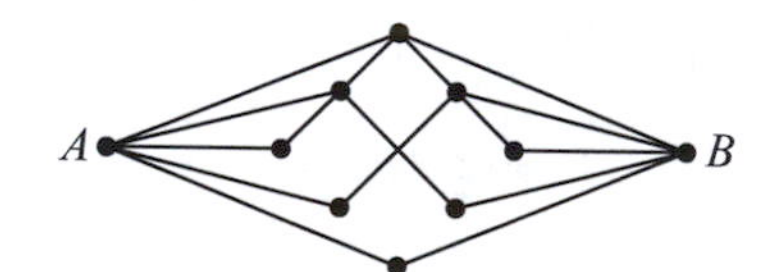

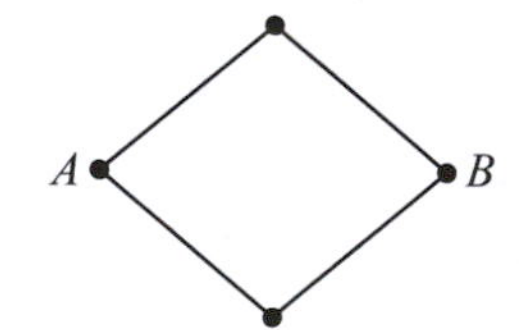

J6 $p(A, B) = 3,\ p'(A, B) = 4$

J7 Both are 4. In general, both are $n - 1$.

J9

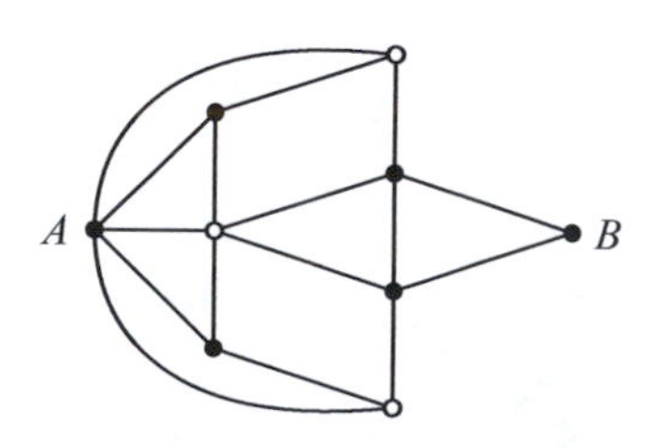

Any path from A to B must include at least one member of the blocking vertex set, and no two internally disjoint paths can contain the same member.

J11

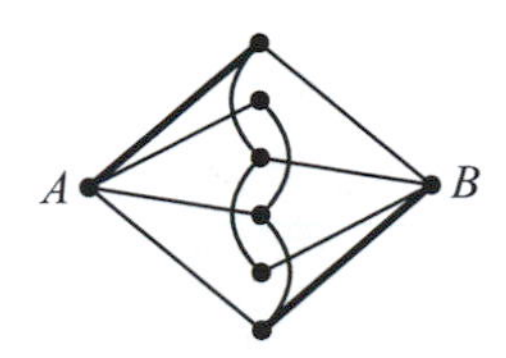

Any path from A to B must include at least one member of the blocking edge set, and no two edge-disjoint paths can contain the same member. Therefore $p'(A, B)$ is at most 2. Two obvious paths show that equality holds.

J13 All of the edges in the graph form a blocking edge set. No blocking vertex set exists relative to A and B if A to B are adjacent.

J14

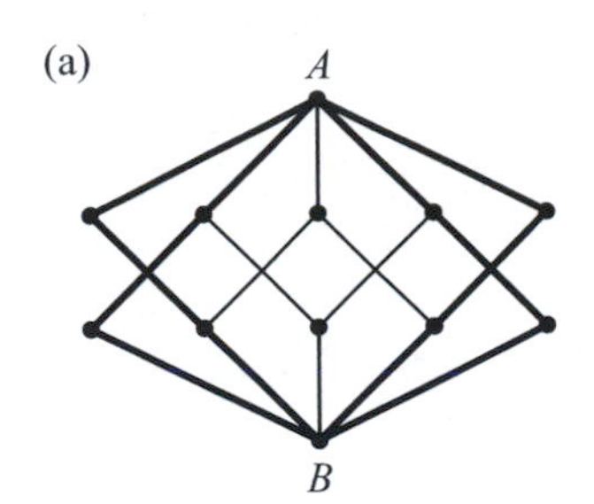

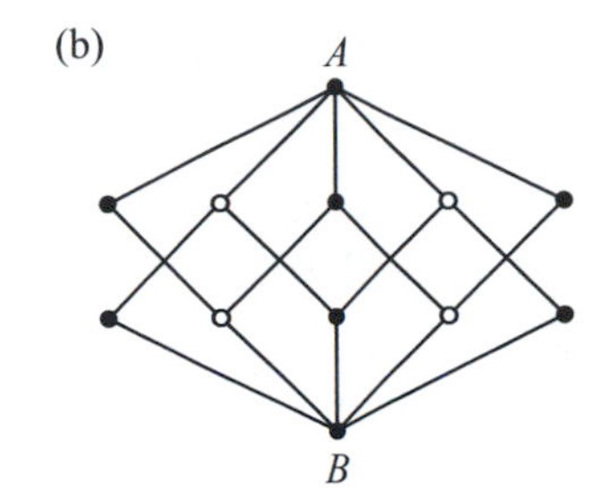

J15 (b) Four edges:

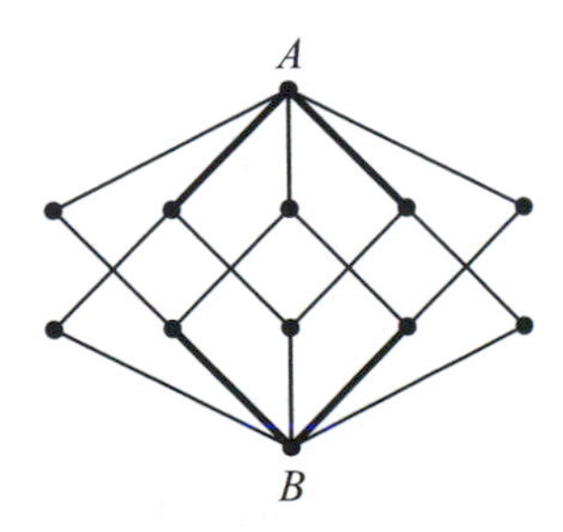

J21 (a) $\delta < n - 1$ since the graph is incomplete.
(b) The vertices adjacent to A form a vertex cut set.

J23 Let **S** be a minimal vertex cut set, and let A and B be vertices in different components of $\mathbf{G} - \mathbf{S}$. **S** is a blocking vertex set relative to A and B, so $p(A, B) \leq q(\mathbf{G})$.

Chapter K

K2 (a) 2 (b) 1

K3 χ only: 5, 3, 2, 3, 3, 4

K4 (a) $\alpha = 3$, and vertices of the same color must be independent.

K5 α, $\alpha\chi$

K9 Color the vertices one at a time, using four colors and always avoiding colors already assigned to adjacent vertices. At each vertex at most three colors have to be avoided, so there is always at least one color available. The same reasoning applies when $\Delta = 100$, using 101 colors. The conclusion is that $\chi \leq 101$.

K10 $\mathbf{K}_5$, $\mathbf{K}_6$, $\mathbf{C}_5$.

K14

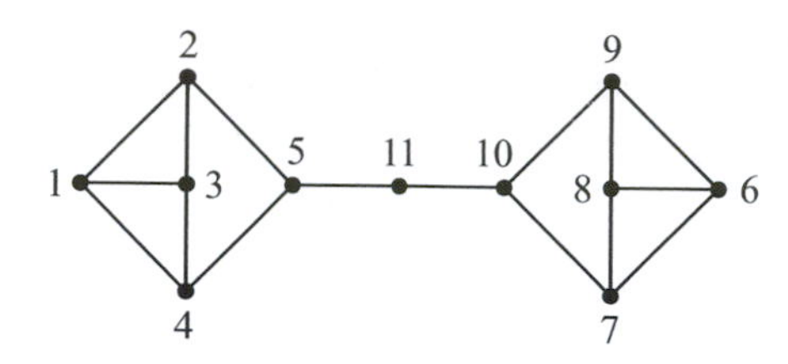

K15 At each vertex at most $\Delta - 1$ adjacent vertices have already been colored, so there is always at least one color available.

K19 Since $d_{k+1} < k$, the only degrees that can be $\geq k$ are $d_1, d_2, \ldots, d_k$.

K20 **H** has fewer vertices than **G**.

K23 If five different colors occur at vertices adjacent to U, then no color is available for U.

K24

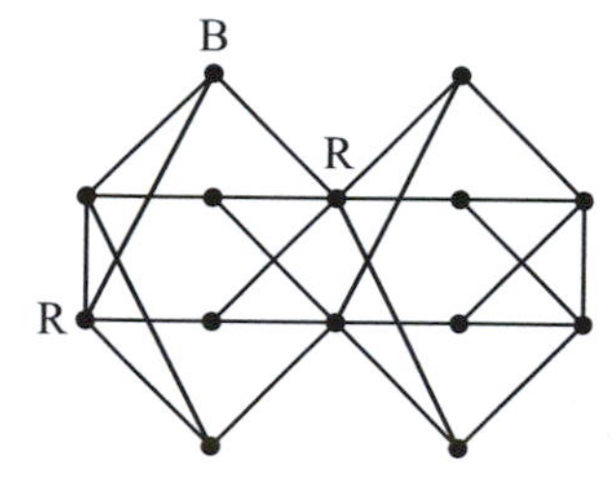

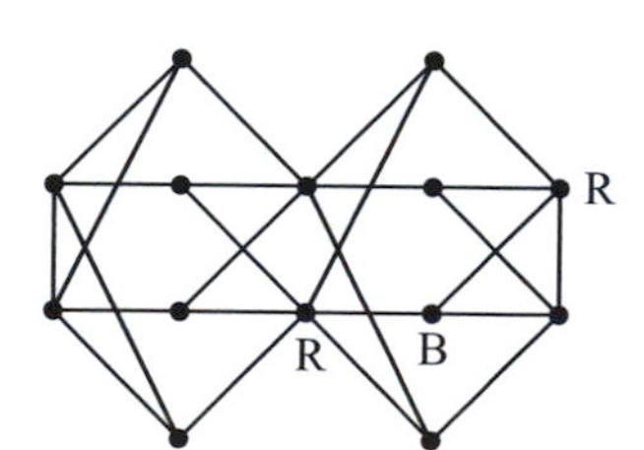

K25 The blue and green vertices, together with their connecting edges, form a connected subgraph.

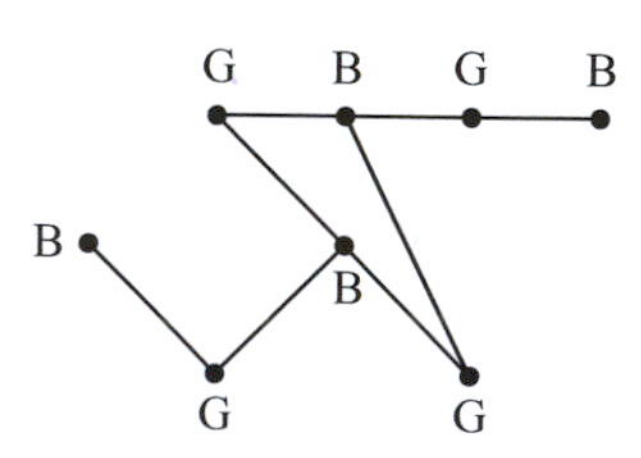

K30 It makes a color available for U.

Chapter L

L1 (a) independent (b) They must all be different.

L3 $\chi' \geq \Delta$

L7 χ' only: (a) 4 (b) 4

L9 $\alpha', \alpha'\chi'$

L10 The number of edges is $n(n-1)/2 = mn$, and $\alpha' = (n-1)/2 = m$. Therefore $\chi' \geq mn/m = n$.

L15 (a) 1 if n is even, 2 if n is odd

(b) 1 if n is even or 1, 2 if n is odd and ≥ 3

(c) 1 (d) 1 (e) 2 (f) 1, 1

L17

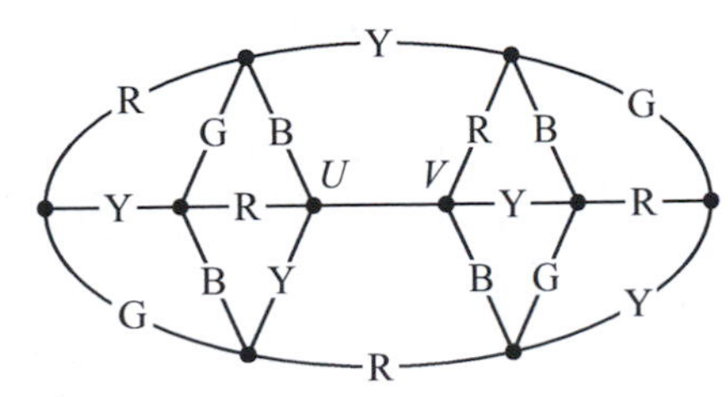

UV can now be green.

L18 (b) A green-yellow switch eliminates either yellow at U or green at V.

L19 There are at most $\Delta - 1$ other edges at that vertex.

L21 If the blue-green switch occurs on an entire path from U to V, then no color becomes available for UV. But such a path would have even length and therefore together with edge UV would form an odd cycle. This cannot occur in a bipartite graph.

Chapter M

M2 (a) $\alpha' = \beta$

(b) $\alpha = \beta'$

(c) $\beta + \beta' = n$

M4 (e) They show that $\alpha' \geq 4$ and $\alpha \geq 7$. But $\alpha + \alpha' \leq n = 11$, so equality must hold.

M6 (b) $\alpha = n - \beta \geq n - k$.

(c) $\alpha + \alpha' \geq (n-k) + k = n$, but also $\alpha + \alpha' \leq n$.

M8 (b) Only upward directions are shown. Black vertices are in **S**.

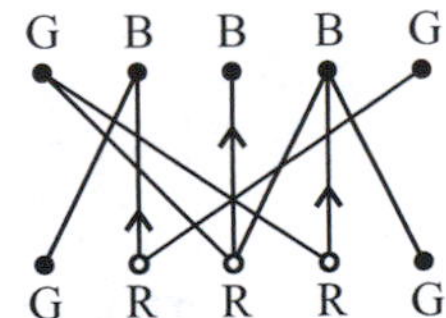

S is not independent.

M9 This path leads to a larger matching, shown at the right.

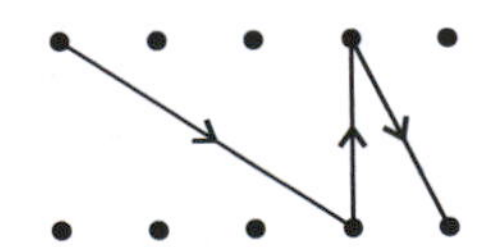

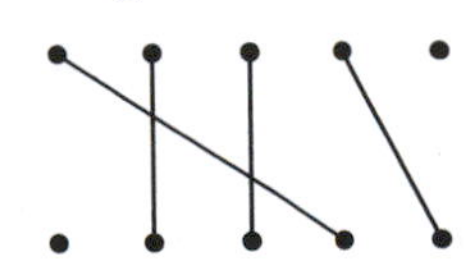

M11 Each edge of **M** is directed from red to blue, so if the red endpoint is reachable from green, then so is the blue endpoint. At the blue endpoint the only incoming edge is the one in **M**, so any path from a green vertex must pass through the red endpoint.

M12 **S** contains all vertices of the graph except for one endpoint from each edge of **M**.

M18 Five independent open cells are indicated.

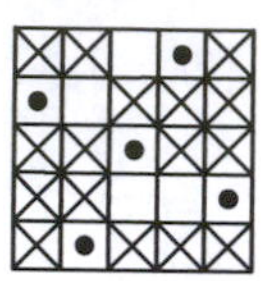

M19 In the first graph below, vertices at the top correspond to rows, vertices at the bottom to columns. In the second graph, vertices on the left correspond to rows, vertices on the right to columns.

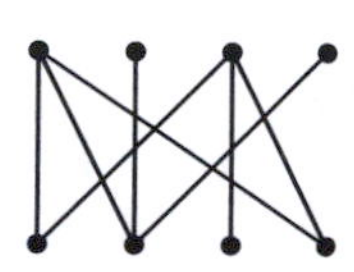

M20 Each edge corresponds to an open cell in the matrix. Each vertex corresponds to a row or column.

M21 An independent edge set corresponds to a set of independent open cells.

M23 The number of cells in a maximal set of independent open cells is equal to the number of lines in a minimal covering line set. A covering line set consists of lines that include all of the open cells in the matrix.

M25 If a matrix of open and closed cells contains a set of k independent open cells, for some number k, and a covering line set consisting of k lines, then the set of independent open cells is maximal and the covering line set is minimal. There is an obvious set of four independent open cells in the example, so this set must be maximal and the line set found in problem M24 must be minimal.

M26 Each row would have to include one of the cells from a set of five independent open cells. But rows 2, 3, and 5 cannot contain more than two independent open cells.

M28 The matrix cannot contain m independent open cells.

M30 The König–Egervary Theorem shows that $r + c$ is equal to the maximal number of independent open cells. By assumption this number is less than m.

M31 (a) Open cells in these rows must be covered by columns in the line set.
(b) c is less than $m + r = k$, violating Hall's condition.

M33 (a) At least kh
(b) At most h
(c) At least k

M34 Corollary 1 applies to (a) and (b). Four independent open cells exist in all three matrices.

Chapter N

N2 The problem is equivalent to finding six independent open cells in this matrix:

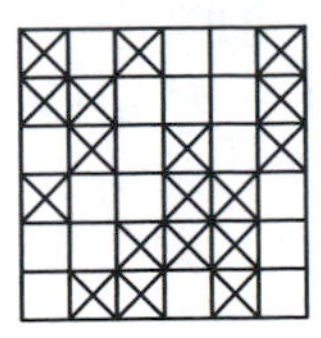

N7 (a) A complete SDR exists for a set of m committees if and only if for each $k = 1, \ldots, m$, the union of any set of k committees contains at least k members.

(b) The union of five committees $(1, 3, 4, 6, 7)$ contains only four members (A, C, E, H).

N12 (a) Each committee corresponds to a row and consists of the elements that are missing from that row. There are m committees, each containing $m - n$ members.

(b) Each element (or person) occurs n times in the matrix, once in each column, and therefore is missing from $m - n$ rows.

(c) Corollary 2 of Hall's Theorem, in committee form, applies to this situation.

N15 (a) r

(b) At most r

(c) The row sums add up to kr. The column sums add up to at most hr. These must be equal, so $k \leq h$.

(d) N contains n independent open cells.

(e) M contains n independent positive entries.

N16 The edges of an r-regular bipartite multigraph can be colored properly using r colors. Each color corresponds to one permutation matrix. This is implied by Edge Coloring Theorem #3, which is valid for bipartite multigraphs.

N17 The total cost of each assignment decreases or increases by the same amount.

N19 (a) Form an n by n matrix of open and closed cells in which open cells correspond to zeros in the cost matrix. The König–Egervary Theorem shows that the maximum number of independent zeros is equal to the minimal number of lines that cover all zeros.

(b) One possibility: rows 1, 3, 5 and column 1.

N22 The positions of n independent zeros determine an optimal assignment.

N24 A strictly decreasing sequence of positive integers must eventually terminate.

N25 n independent zeros must exist when the algorithm terminates. Otherwise all zeros would be covered by fewer than n lines and the algorithm would continue.

Chapter O

O3 $\{00, 01, 02, 03, 04, 14, 24, 34, 44\}$ is a maximal chain.
$\{04, 13, 22, 31, 40\}$ is a maximal antichain.

O4 (a) A is a subset of B.

(b) $\{\varnothing, 1, 12, 123, 1234\}$ is a chain.
$\{12, 13, 14, 23, 24, 34\}$ is an antichain.

O5 (a) A is a divisor of B.

O6 $\{ABGH, EFCD, I\}$

O8 k chains. Each member of the antichain must be in a different chain.

O9 In a cycle-free digraph, if an antichain contains k members and a chain decomposition contains k chains, for some number k, then the antichain is maximal and the chain decomposition is minimal.

O13 $\{12, 13, 14, 23, 24, 34\}$ is a maximal antichain.
$\{(\varnothing, 1, 12, 123, 1234), (2, 23, 234), (3, 13, 134), (4, 14, 124), (24), (34)\}$ is a minimal chain decomposition.

O16 A minimal chain decomposition contains m_1 chains. Each chain contains at most m_2 elements. Every vertex in the graph is in one of these chains, so there are at most $m_1 m_2$ vertices.

O17

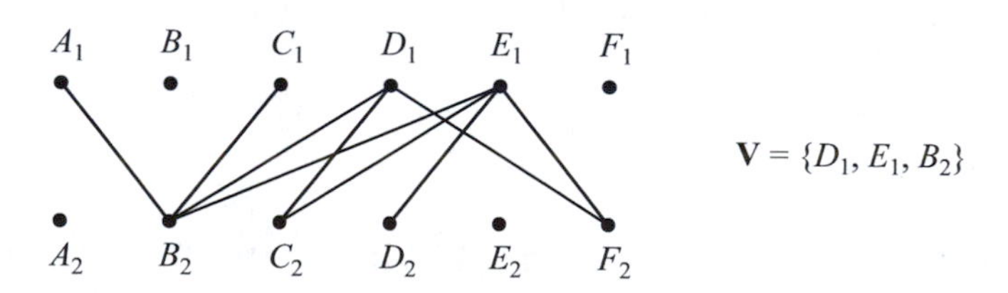

(a) $\{A_1B_2, D_1C_2, E_1D_2\}$ is an independent edge set, so this edge set is maximal and $\mathbf{V}$ is minimal.

(b) (A, C) is an antichain.

O19 $\mathbf{V}$ contains β vertices, which correspond to at most β different vertices in $\mathbf{D}$. The remaining vertices in $\mathbf{D}$, at least $n - \beta$, form an antichain.

O20 (a) Using the matching $\{C_1B_2, D_1C_2, E_1D_2\}$:

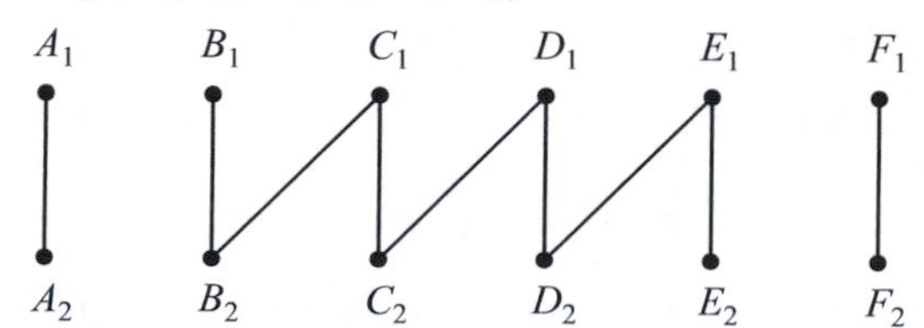

The chain decomposition is $\{A, EDCB, F\}$.

(b) Using the matching $\{A_1 B_2, D_1 F_2, E_1 C_2\}$:

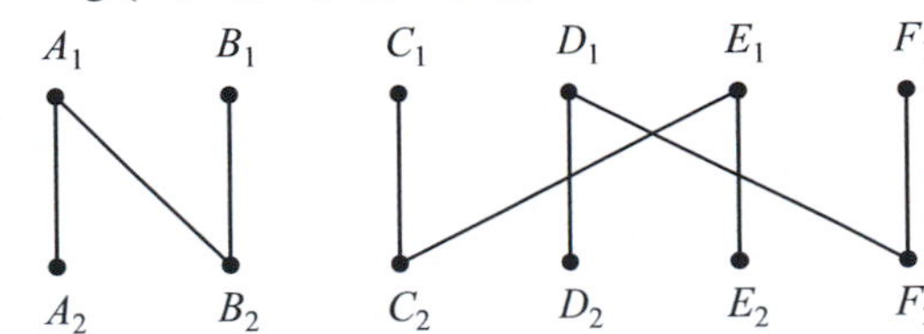

The chain decomposition is $\{AB, EC, DF\}$.

O21 (b) (a) implies that each component of $\mathbf{B}^*$ is either a simple path or a cycle. But a cycle in $\mathbf{B}^*$ would correspond to a cycle in $\mathbf{D}$, which is impossible.

O22 (a) $\mathbf{B}^*$ contains $2\alpha'$ vertices of degree 2, corresponding to the vertices covered by a maximal matching in $\mathbf{B}$. The remaining $2n - 2\alpha'$ vertices in $\mathbf{B}^*$ have degree 1.

(b) Each component has two endpoints of degree 1, so the number of components is $(2n - 2\alpha')/2 = n - \alpha'$.

O25 The König–Egervary Theorem shows that $\beta = \alpha'$, so $n - \beta = n - \alpha'$. An antichain cannot have order greater than that of a chain decomposition, so the antichain must have order exactly $n - \beta$. By the max-min principle, the antichain is maximal and the chain decomposition is minimal.

Index

α, 85
α', 85
β, 89
β', 89
$c(AB)$, 164
D, 133
$\Delta(\mathbf{G})$, 14
$\delta(\mathbf{G})$, 14
$\mathbf{K}_5$, 80
$\mathbf{K}_n$, 11
$\mathbf{K}_{m,n}$, 26
$q(\mathbf{G})$, 99
S, 87
S$'$, 87
ϕ, 161
$\phi(AB)$, 161
$v(\phi)$, 163
χ, 103

adjacency matrix, 16
adjacent vertices, 1, 9
antichain, 154
antichain, order, 154
arcs, 161

bipartite complement of **G**, 29
bipartite graph, 25, 66
blocking edge set, 97
blocking vertex set, 96
Bondy–Chvatal condition, 69
Bondy–Chvatal Theorem, 69, 72
Brooks' Theorem, 105, 115

capacity of a cut, 169
capacity of the arc, 164
Cayley's Formula, 34
chain, 153
chain decomposition, 154
    length, 153
    order, 154
chain in a digraph, 153
clique, 92
clique number, 92
closed Euler path, 57, 58
closed path, 12
codeword, 35
    long, 36
    short, 36
color switch, 109
colored digraph **D**, 133
Colored Digraph Theorem, 135
complement of a graph, 13
complete bipartite graphs, 26
complete graph, 11
complete SDR, 143
complete system of distinct representatives, 143
components of a graph, 15
connected graph, 15
constructing spanning trees
    building up, 41, 42
    reducing down, 41, 42
covered vertices, 133
covering edge set, 89
covering numbers, 89
Covering Theorem #1, 90, 93
Covering Theorem #2, 90, 94
covering vertex set, 89
cut, 168
    backward arc, 169
    forward arc, 169

cycle, 12
cycle flow, 162
cycle graphs, 11
Cycle Theorem for Bipartite Graphs, 27, 28
cycle-free digraph, 153

DeBruijn sequence, 61
Decomposition theorem for integral flows, 163
degree of a vertex, 13
Degree Sequence Algorithm, 17
degree sequence of a graph, 14
Degree Theorem, 14
demand, 175
demand nodes, 175
diameter of a graph, 19
digraph, 11
Dijkstra's Algorithm, 46, 53
Dilworth's Theorem, 155
Dirac's Theorem, 68
directed graph, 11
directed tree, 33, 44
disconnected graph, 15
distance (between two vertices), 19
dodecahedron, 80

edge, 1, 9
edge color switch, 122
edge coloring of a bipartite graph, 122
edge coloring number, 119
Edge Coloring Theorem #1, 121
Edge Coloring Theorem #2, 121, 127
Edge Coloring Theorem #3, 122, 123
edge cut number, 101
edge cut set, 101
edge disjoint paths, 6
efficient, 176
equivalent graphs, 9
Euler path, 2, 57
  closed, 57
  open, 57
Euler Path Theorem, 58
Euler's Formula, 5, 79, 80
even vertex, 14
extended maximal flow algorithm, 182

finite graphs, 9
Five Color Theorem, proof, 108
flow, 161
  cycle, 162
  equivalent, 164, 182
  feasible, 165
  integral, 163
  N-reduced, 182
  reduced, 164
flow conservation equations, 161
Ford's Algorithm, 49, 54
forest, 31
Four Color Theorem, 2, 108

Gale's Feasibility Theorem, 179
graph, 9
  2-colorable, 25
  bipartite, 25
  components, 15
  connected, 15
  disconnected, 15
  forest, 31
  mixed, 11
  non-planar, 81
  tree, 31
  weighted, 41
  complete bipartite, 26
  isomorphic, 21
  regular, 15
graph game, 88

Hall's condition, 138
Hall's Theorem, 138
Hamilton path, 2, 65
Heawood, Percy, 115
height of a vertex, 159
Hungarian Algorithm, 148
Hungarian step, 148

icosahedron, 80
improvement step, 49
incidence matrix, 16
indegree, 33
independence numbers, 85
Independence Theorem #1, 86
Independence Theorem #2, 87
Independence Theorem #3, 87
independent edge set, 85
independent vertex set, 85
integral flow, 163
internally disjoint paths, 6, 95
isomorphic graphs, 4, 21

k-connected, 98
k-edge-connected graph, 102
König's Coloring Theorem, 122
König–Egervary Theorem, 87, 133, 136
Königsberg Bridge Problem, 1, 57
Kempe, Alfred, 115
Kuratowski's Theorem, 82

labeled graphs, 10

latin square, 144
Lemma on Vertices of Degree 1, 32
length of a chain, 153
length of a path, 12
line in a matrix, 136
long codeword, 36
longest simple path, 12
loops, 60
lower capacity problem, 179

Map Coloring Theorem, 110
matching, 132
matching extension algorithm, 135
matrix of open and closed cells, 136
max/min principle, 132
maxflow/mincut theorem, 171
maximal flow algorithm, 165, 167
maximal independent edge set, 85
maximal independent set, 85
maximum degree, 14
Menger's Theorem, 97
Menger's Theorem, corollary, 99
minimal covering edge set, 89
minimal covering sets, 89
minimal covering vertex set, 89
minimal cut algorithm, 171
minimal path, 45
Minimal Path Algorithm, first attempt, 45
Minimal Path Algorithm, revised, 46
minimal path problem, 45
minimal proper edge coloring, 119
minimal proper vertex coloring, 103
minimal spanning tree, 42
minimum degree, 14
mixed graph, 11
monotone sequence, 7
multigraph, 10
multiple edges, 10

negative cycle, 50
negative test for bipartite graphs, 66
negative weight, 48
network, 161
new demand, 178
node capacities, 173
nodes, 161
    demand, 175
    supply, 175
non-planar graph, 81

odd closed path, 27
odd vertex, 14
open Euler path, 58
open path, 12
optimal assignment problem, 146
optimal spanning tree, 46
order of a chain decomposition, 154
order of an antichain, 154
Ore's adjacency condition, 75
Ore's Theorem, 75

partial latin square, 144
path, 11
    edge-disjoint, 95
    internally disjoint, 95
path connection numbers, 96
path flow, 162
Path/Cycle Principle, 69
perfect matching, 88
permutation matrix, 145
planar graph, 4, 77
plane diagram, 77, 78
Posa's degree conditions, 68
Posa's Theorem, 68, 76
Prim's Algorithm, 42, 53
    tables, 43
proper edge coloring, 3, 119
proper vertex coloring, 103
Prufer's Method, 35
pruning a tree, 32
Pruning Lemma, 32

reachable (vertex), 33
reduction algorithm, 43, 53
Regional Degree Theorem, 79
regional degrees, 79
regions in a plane diagram, 78
regular graphs, 15
root, 33, 44
Root Theorem, 34

saturating path flow, 165
Scheduling Problem, 124
short codeword, 36
shortest path, 12
shortest path problem, 44
simple path, 12
sinks, 161
Six Color Theorem, proof, 108
sources, 161
spanning tree, 34, 44
state diagram, 60
subdivision of a graph, 81
subgraph, 13, 67
Subgraph Test, 67
supply, 175

supply nodes, 175
system of distinct representatives (SDR), 143

tree, 31
Tree Theorem 1, 32
Tree Theorem 2, 33

unit path flow, 162
unlabeled graphs, 10
unlabeled trees, 31
Upper Bound Algorithm for χ, 107

value of ϕ, 163
vertex, 1, 9
vertex coloring number, 103
Vertex Coloring Theorem #1, 105
Vertex Coloring Theorem #2, 105
Vertex Coloring Theorem #3, 107
Vertex Coloring Theorem #4, 108
vertex cut number, 99
vertex cut set, 99, 115
Vizing class 1, 122
Vizing class 2, 122
Vizing-Gupta Theorem, 121

weight, 41
weight of a path, 45
weight of a tree, 42
weight, negative, 48
weighted graph, 41

About the Author

Daniel A. Marcus received his PhD from Harvard University. He was a J. Willard Gibbs Instructor at Yale University from 1972–74 and Professor of Mathematics at California State Polytechnic University, Pomona from 1979–2004. Marcus has published research papers in the areas of graph theory, number theory, and combinatorics. He is the author of the following books: *Combinatorics: A Problem Oriented Approach* (also with the MAA), *Differential Equations: An Introduction,* and *Number Fields*.